Microsoft .NET Gadgeteer
Electronics Projects for Hobbyists and Inventors

Simon Taylor

New York   Chicago   San Francisco
Lisbon   London   Madrid   Mexico City
Milan   New Delhi   San Juan
Seoul   Singapore   Sydney   Toronto

The **McGraw·Hill** Companies

Cataloging-in-Publication Data is on file with the Library of Congress

McGraw-Hill books are available at special quantity discounts to use as premiums and sales promotions, or for use in corporate training programs. To contact a representative, please e-mail us at bulksales@mcgraw-hill.com.

Microsoft .NET Gadgeteer: Electronics Projects for Hobbyists and Inventors

1 2 3 4 5 6 7 8 9 0   QFR QFR   1 0 9 8 7 6 5 4 3 2

ISBN 978-0-07-179795-5
MHID   0-07-179795-5

Sponsoring Editor Roger Stewart	**Copy Editor** Lisa Theobald	**Composition** TypeWriting
Editorial Supervisor Janet Walden	**Proofreader** Paul Tyler	**Art Director, Cover** Jeff Weeks
Project Editor Patricia Wallenburg	**Indexer** Jack Lewis	**Cover Design** Jeff Weeks
Acquisitions Coordinator Molly Wyand	**Production Supervisor** James Kussow	

To my wife, Tsvetana, whose support and
encouragement keeps me going.

About the Author

Simon Taylor (Southampton, United Kingdom) has been working with embedded device hardware and software for more than 30 years. In 1994 he created the design consultancy Sytech Designs Ltd., which he still runs today. His career started as an Electronics engineer with the Royal Air Force, specializing in air radio and navigation. After leaving the Air Force, he started working in the newly emerging embedded computer world, initially working on ground-breaking digital video "framestores" and special effects units, leading to his being head-hunted by a Silicon Valley video technology company and relocating to sunny California.

This was followed by a move to the "space town," Huntsville, Alabama, where he worked for a theatrical lighting manufacturer. In addition to lighting and laser controllers, he designed several serial control protocols and a system of recording lighting control signals on VHS video tape alongside the stereo audio. Following his return to the United Kingdom, he started Sytech Designs. Sytech Designs specializes in embedded designs, machine-to-machine communications, and transport systems. In his role at Sytech Designs, he has led design projects involving communications, GPS, GSM, and card payment systems. He has been involved with Micro Framework since the early days and is a member of the Micro Framework Core Team and is a Micro Framework partner. Sytech Designs is one of the first manufacturers designing and manufacturing Gadgeteer mainboards and modules.

Contents

Foreword

Frederick Brooks, the author of the seminal book *The Mythical Man-Month: Essays on Software Engineering* (Addison-Wesley Professional), described the delights of programming as including the "sheer joy of making things" and "the delight of working in such a tactile medium" where "it prints results, draws pictures, produces sounds, and moves arms." Almost every programmer I speak with remembers their early experiments with LEDs and servos, even if they have been away from building devices for some time.

I have had the opportunity to teach programming at several levels, and I have seen that creating things that interact with you, with other objects, and with the environment around them can be very engaging. There is also the opportunity for experimenting in the explosion of connected devices that will instrument our world in the next generation. That is what .NET Gadgeteer is all about.

The challenge in approaching these kinds of projects has always been the steep initial learning curve. In the past, to complete an electronics project even as simple as turning on an LED, you needed to know where to get and how to select compatible electronic components, load compatible development tools, learn a new language or a new dialect of a desktop language, pick up some fabrication skills such as breadboarding or soldering, learn embedded interfaces at least to the level of GPIO, find out how to deploy your code to the device, and then heaven help you if it didn't do what you wanted because you had to figure out how to debug what you put together (wiring, logic, and all).

.NET Gadgeteer reverses this equation so that you don't have to learn a lot before you can create something simple—now you can create something compelling very quickly, and you can drill into the details as you need and want to. The impact of this is evident. Nearly every day, new blog entries are written by someone who has taken Gadgeteer off the shelf (so to speak) and created something cool very quickly. For example, the standard first demo that we do on the NETMF/Gadgeteer team is to write a functional camera in about four lines of code with components that just plug together. This simplicity is important, because it means that the reinforcement of building cool stuff starts immediately.

However, once you have built a simple camera or gotten a servo to go where you want it to, what do you do next? The quick four-line camera demo takes a picture when you press a button and displays the image on the screen. The next step might be to save that picture to an SD card and then be able to scroll through the images on the SD card and delete those that are not as good as you first thought. Suddenly, you start to get into more than four lines of code, and things such as FileIO and a touch UI and possibly more.

This is why I really like this book: It takes you beyond the simple projects—many of which you can find on the Internet—and introduces some of the underlying information that you will need if you keep going past that first project and begin to create projects of your own. For example, you can build lots of cool stuff without knowing what I2C is, but at some point you may want to use a particular sensor that is accessible only through I2C. I rarely create a project these days that does not include an extender module—one that exposes the underlying functionality of the micro-controller unit (MCU). This is because there is not always a module for what I want to do, or the module that exists does not use exactly the right components that I need. This book offers an easy entry to .NET Gadgeteer and the world of making things that work, and it also gives you the background that you will need as you continue on—to keep going.

The topics covered in depth in this book include working with Visual Studio, including how it is the same as and different from building other applications in that tool set; how to manage power when your device is running on batteries; and what the event-driven design model is and why it is the core of the Gadgeteer applications model. These are just a few examples of the many areas covered here. So if you have copied some fun applications from the Web and now you want to either extend them or start new ones of your own, this book will give you the understanding that you will need to go almost anywhere in building out small devices.

Colin Miller
Microsoft Product Unit Manager,
.NET Fundamentals

Acknowledgments

I want to thank the .NET Micro Framework team at Microsoft, in particular Colin Miller, Lorenzo Tessiore, and Zach Libby, for all their help over the last few years.

Thanks are due to the .NET Gadgeteer team at Microsoft Research, Cambridge: Nicolas Villar, James Scott, and Steven Johnson for their highly infectious enthusiasm for .NET Gadgeteer.

A special thanks to all the editing staff at McGraw-Hill Professional for their work in making this book happen.

Finally, a special thanks to my wife, Tsvetana, whose encouragement and support made this book possible.

Introduction

An embedded device is a combination of hardware and software. The hardware requires specific sensors and controls to interface with the real world. The software defines the behavior of the device and how it reacts to the sensors and controls.

Microsoft .NET Gadgeteer defines a hardware and firmware standard, allowing manufacturers to design a range of sensor boards that can be plugged into compatible processor boards. This simplifies the first requirement of an embedded device: hardware. You can select suitable sensors and interfaces that are required for your project and just plug them into a processor board. The standard defines the interface required by a processor board, allowing you to use a range of processor boards from different manufacturers. .NET Gadgeteer also defines the firmware interface between the sensors and the mainboard, allowing the sensors to be provided with a low-level driver, used by the processor board.

The second requirement is the software that implements the functionality of your application. The Microsoft .NET Micro Framework provides a runtime operating system with a programming interface based on the PC desktop .NET Framework. A complete development and debugging tool set is also provided, in the form of Visual Studio Express. Your application is written in a high-level language (C# or Visual Basic).

The combination of Gadgeteer-compatible hardware and the .NET Micro Framework provides a simple system of assembling custom hardware and implementing complex embedded software.

The purpose of this book is to explain how all the elements in .NET Gadgeteer work, so you have the knowledge to create your own projects. We will explore not just how to use various Gadgeteer sensors in projects, but the underlying principles of .NET Gadgeteer and different aspects of programming. Your understanding of how .NET Gadgeteer works and interacts with .NET Micro Framework will make developing a bug-free complex application much simpler.

This book covers all aspects of using .NET Gadgeteer, from setting up the development environment, to designing your application and debugging techniques. We use example projects to illustrate various programming techniques and aspects of .NET Gadgeteer.

By the end of the book, you should have the knowledge and confidence to tackle almost any small, embedded device project.

PART I

Getting to Know
.NET Gadgeteer

CHAPTER 1

Introduction to .NET Gadgeteer

M icrosoft .NET Gadgeteer was developed by Microsoft Research Cambridge. It started as an internal project intended to simplify the development of embedded devices. The aim was to design a standardized mechanism for developing hardware sensors and interfacing them to .NET Micro Framework–based hardware. The wider usefulness of this project soon became apparent, however, and it was decided to make it public and change it to an open source project.

NOTE *In this book, I will refer to the Microsoft .NET Gadgeteer (a registered trademark owned by Microsoft) product in the shortened form, "Gadgeteer."*

.NET Gadgeteer Basics

Gadgeteer is a combined physical hardware interface and a firmware framework that serves as a standardized mechanism, or toolkit, for connecting hardware sensors and peripherals (called *modules*) to a Micro Framework–based processor board (called a *mainboard*). Gadgeteer allows modules designed by any manufacturer to be connected to mainboards from any manufacturer and simplifies integration into a user application. We will refer to the Gagdeteer framework as the Gadgeteer core, to avoid any confusion with the Micro Framework runtime it sits on.

Hardware Interface

In general, Gadgeteer mainboards are black, and modules that supply power are differentiated by the color red. Gadgeteer's physical hardware interface controls the interconnection between modules and mainboards. It defines the type of connection headers used (10-pin, 1.27mm headers) and the pin functions of the connectors (which pins are for data and which are for power). The physical connection is made using polarized insulation-displacement connectors (IDCs) and ribbon cables. The physical board properties are also defined, including mounting hole locations and pitch, as are connector position and board sizes.

Firmware

The Gadgeteer core firmware implements a number of generic hardware-interface functions. These range from digital inputs and outputs, to serial connections, to hardware communications protocols such as Inter-Integrated Circuit (I2C) and Serial Peripheral Interface (SPI). These hardware interfaces use the Microsoft .NET Micro Framework runtime to connect to physical processor hardware. Module firmware drivers use these interfaces in their low-level drivers, allowing connection to the mainboard hardware using the Micro Framework.

The modules implement a firmware Gadgeteer driver that encapsulates the hardware interface and connection. A simple high-level application programming interface (API) is exposed, allowing an application to use the functions of a module, without your needing to worry about the details of the low-level code (which is provided for you). For instance, a temperature sensor could expose a property called **CurrentTemperature**, which, when called, returns an integer of the current temperature in degrees centigrade. You do not need to worry about the details of how the temperature sensor works or how to interface to it from a low-level perspective; you simply need to get the temperature for use in your application.

Micro Framework and Gadgeteer

The Gadgeteer core sits on top of the Micro Framework, which allows the embedded application to be written in a high-level language using Microsoft .NET.

The Micro Framework is a version of the .NET Framework, written for limited resource 32-bit ARM processors. It allows application development in managed code, using free Visual Studio tools (Express edition) and C#. Visual Studio supplies a complete development environment, allowing you to deploy applications and allows hardware debugging. Gadgeteer even installs a *graphical designer* into Visual Studio, allowing you to drag-and-drop representations of modules and mainboards from a toolbox onto the designer interface. You can then add connections from the modules to the mainboards and generate application code stubs for the project.

Micro Framework is an implementation of the .NET Common Language Runtime (CLR), which was written especially for 32-bit ARM microprocessors. It is not a scaled-down version of the desktop CLR (as is Windows CE .NET Compact Framework), but a complete rewrite of a subset of the .NET Framework for ARM processors, which takes into account the limited memory and resources of an embedded processor. The Framework is written in native code using C and C++ and optimized for reduced instruction set computing (RISC) processors.

Currently the Gadgeteer core runs on version 4.1 of the .NET Micro Framework, and the application programming language is C#. When the Gadgeteer core supports Micro Framework 4.2, Visual Basic .NET will be included as an additional programming language. The Micro Framework provides a subset of the desktop .NET API. Additional API functions not supported in desktop .NET are specific to an embedded development environment, such as I2C and SPI hardware communication functions.

The complete Micro Framework architecture is shown in Figure 1-1. Micro Framework is a scaled-down version of the CLR, called the *TinyCLR*, that sits on top of the hardware to make a bootable runtime. Micro Framework serves as a buffer between the hardware and the application code, providing a common programming interface between the different hardware platforms.

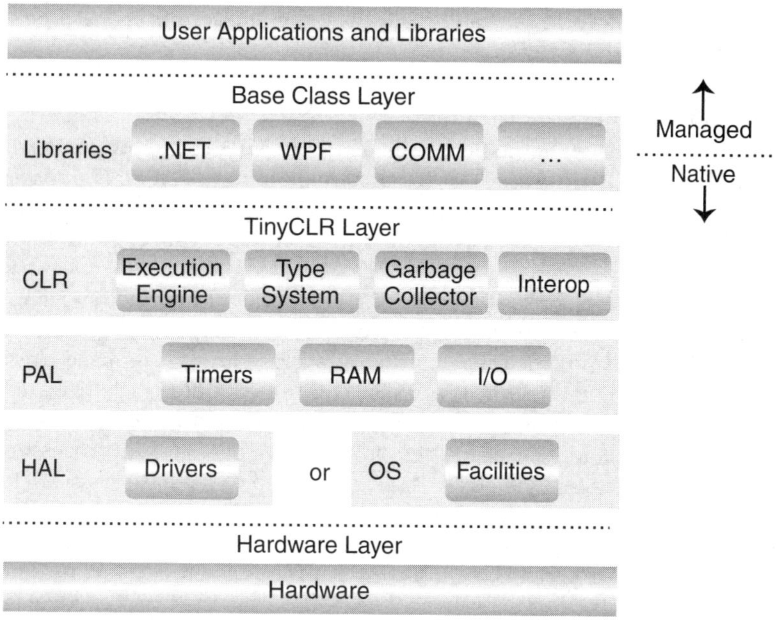

FIGURE 1-1   Micro Framework architecture

The Micro Framework comprises three main layers: the Hardware Layer, the TinyCLR Layer, and the Base Class Layer.

TinyCLR

The foundation of the TinyCLR is the abstraction to the physical hardware. The HAL (Hardware Abstraction Layer) provides a common access to the actual hardware elements. This needs to be customized for each different hardware platform. It is possible for this to be implemented to use an underlying OS (operating system), rather than direct access to hardware. This is the case for the "Emulators" that run on a PC using Windows. The PAL (Platform Access Layer) uses the HAL interface and generally does not need customization between hardware platforms. The HAL provides elements such as Timers, RAM usage, and general high-level I/O functions. The CLR provides the managed execution engine, managed types, Interop, and the Garbage Collector.

The Execution engine runs the IL (Intermediate Language) code, produced by the Visual Studio build process. The various elements in the CLR layer provide the Micro Framework features such as "time sliced threading management" and exception handling.

The Garbage Collector provides memory management of variables; it allocates memory to variables and monitors the application usage of these variables. When the variable is no longer in use by the application, the Garbage Collector will free the memory allocated and allow it to be reused.

Interop allows custom extensions of the Micro Framework. Additional functions can be added to the code, using native code (C/C++ code) and an interface to manage access by the managed (C#) code layer. Adding an Interop extension requires access to the source code for the port and a supported compiler/linker, as well as an in-depth knowledge of the porting process.

Base Class Layer

The Base Class Layer is the highest level of the Micro Framework Port. This provides the API to the .NET libraries, with functionality such as WPF (Windows Presentation Foundation—graphics components), serial comm ports, network sockets, etc.

Above the Base Class Layer are the User Applications and Libraries, such as Gadgeteer.

The C# compiler generates processor-independent intermediate language (IL) code that the TinyCLR executes on the device.

The TinyCLR abstracts the hardware access through its base class libraries and treats hardware components as objects. This allows different hardware to be accessed the same way from an application's point of view. Different hardware platforms (for instance, a Cortex M3–based processor and an ARM 7–based processor) have custom hardware abstraction layers (HALs) implemented to the base classes. This HAL implementation is the main task involved in porting a hardware platform to Micro Framework. This work is normally done by the manufacturer of the Micro Framework hardware. However, the Micro Framework Porting Kit is open source, so you can use it to create your own custom port of a hardware base (but this is an advanced level task). A number of open source platform ports can be used as the basis for your own custom port.

Gadgeteer Architecture

Gadgeteer sits between the Micro Framework and the user application, as shown in Figure 1-2. The Gadgeteer system defines the physical hardware interface between modules and mainboards and supplies a software framework, allowing simple interfacing and integration.

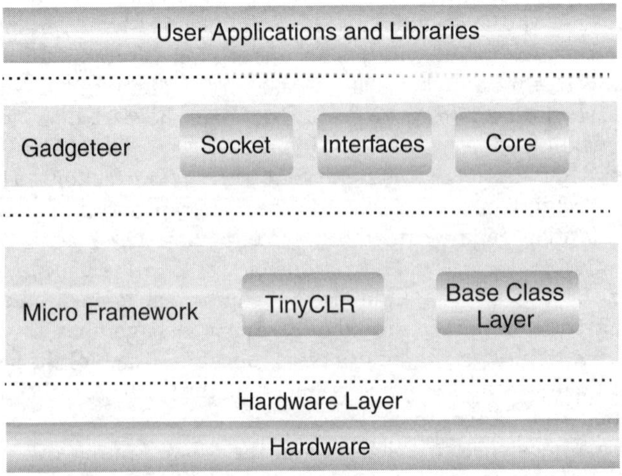

FIGURE 1-2 Gadgeteer and Micro Framework

Hardware Interface

The key hardware element of Gadgeteer is the physical connection between mainboards and modules. This is defined as a 10-pin, 1.27mm header, with an IDC ribbon cable. The 1.27mm header was chosen for its combination of small size, robustness, and ability to be polarized; the cable cannot be plugged in the wrong way (unless you apply extreme force using a small hammer!).

The pin-out of the headers is also designed to simplify use. Pins 1 and 2 supply power (+3V3 and +5V), and ground is on pin 10. Pins 3 through 9 are defined as data pins. As the connector and cables are polarized, power can never be applied to the wrong pins of a module or mainboard. If you plug a module into the wrong mainboard connector, at the worst only data pins will be mismatched to the wrong data pin functions, so no catastrophic failures should occur—that is, you shouldn't accidentally fry your mainboard! The interconnection method is not designed to allow hot-plugging, however.

CAUTION   *Connect and disconnect modules to a main board only with the power turned off.*

The connector orientation on the boards is also defined. The connectors are positioned on the polarizing slot to allow the ribbon cables to be plugged in from a mainboard to a module without twists in the cable. In addition, all Gadgeteer boards are recommended to have radiused (rounded) corners—with a mounting hole in each corner. The pitch of mounting holes should be on a 5mm grid, with the holes being 3.5mm from the edge of the boards.

The Gadgeteer framework defines a number of hardware functions, which are referred to as *interfaces* in Gadgeteer. The following is a list of the hardware interfaces supported:

- General purpose input/output (GPIO), with or without interrupt capability
- Serial universal asynchronous receiver/transmitter (UART), with or without handshake
- I2C bus
- SPI bus
- Analog inputs
- Analog outputs
- USB host
- USB device
- Controller-area network (CAN) bus
- Graphics display
- Touchscreen
- Pulse Width Modulation (PWM)
- Ethernet
- Secure Digital (SD) card
- Manufacturer-specific interfaces

The link between modules and mainboards is called a *socket*. A mainboard physical socket can support one or more of these interfaces. Gadgeteer defines the pin-out for each interface function on a socket. The socket pin definitions are defined so it is possible for a socket to support multiple functions—for instance, a socket could be defined as supporting serial UART and GPIO, being used as a serial connection in one application and as GPIO in another.

Each interface function is represented by a letter to identify it. For instance, "I" represents I2C and "S" represents SPI. Each physical socket is labeled with all the interface letters supported on that socket. A module socket is labeled with the type of mainboard interface socket required—for instance, a module with a SPI interface to the mainboard would have its socket labeled "S" and would need to be connected to a mainboard socket labeled "S."

A module manufacturer supplies a peripheral/sensor, which requires a specific socket type. A module can require more than one socket type—for instance, a graphics display with a touchscreen will require four sockets: red, green, and blue display sockets and a touch socket.

Firmware Interface

The Gadgeteer core defines a base class for mainboards and modules. Each Gadgeteer-supported hardware function is implemented as a class from a library referred to as an interface. A module's firmware uses one or more of the hardware interfaces and associates them with a socket, or in some cases multiple sockets.

A mainboard manufacturer designs a mainboard with a number of physical sockets. Each socket will support one or more functions. The mainboard firmware will implement a model of the hardware capabilities, inheriting from the Gadgeteer mainboard class. The mainboard firmware will define all the sockets used, connecting each pin to the physical hardware pin and defining the supported socket functions. It will then add these sockets to the Gadgeteer core socket collection, creating a model of the physical connections and functions available for this mainboard. Note a mainboard will not necessarily support all the Gadgeteer-supported interface functions, just the ones supported by that particular mainboard.

A module manufacturer supplies a peripheral/sensor that requires a specific hardware socket type. Then the firmware for the module is implemented, inheriting from the Gadgeteer module class.

The module manufacturer writes the low-level driver to use the relevant Gadgeteer interface class, such as a GPIO input for a pushbutton. The module also implements the low-level driver code for the hardware functions using the Gadgeteer interface.

Modules (peripherals) include a software model of the hardware interface required, and mainboards (processors) include a software model of their physical hardware capabilities and the physical connections exposed.

Creating Gadgeteer Applications

The remaining part of the picture is a Gadgeteer *application*. This is provided by the Gadgeteer core, which creates the instances of the required module classes and the instance of the mainboard class, connects the modules to the mainboard, and starts the Micro Framework application. All the user needs to do is write his or her application functionality that uses the modules; all the background setup and low-level code has been completed for them.

Serial Camera Module

We will now examine a real-world example of a Gadgeteer application using a serial camera module. The module is a camera that can take pictures and return them as JPEG files. The camera is controlled and configured using a serial protocol. The JPEG pictures are then downloaded from the camera over the serial connection.

The manufacturer of the serial camera implements a firmware driver that handles the low-level serial protocol, configures the camera, sends commands to the camera, and downloads the JPEG pictures. A Gadgeteer serial interface class is used to achieve this. A high-level API is provided for application users, which includes simple functions such as **TakePicture**. The picture-handling process occurs on a background thread, so as not to hold up the main application. When the picture is ready, a **PictureReady** event is fired, passing back the JPEG picture in the event. All the details of communicating with the camera are hidden from the user.

The user simply uses a ribbon cable to plug the serial camera module into a socket on the mainboard that supports the serial interface. This connection provides the serial data link to the module and also supplies the required power supply to the module. The module firmware library is then added to the project; the mainboard firmware defines that particular socket as supporting a serial interface. The Gadgeteer framework will connect the module driver to the mainboard socket in the Gadgeteer user application. The mainboard firmware uses the Gadgeteer core to interface the serial socket, and the module firmware implements the code to interface the sensor functionality to the Gadgeteer core.

In the application, an instance of a camera object exists and has been interfaced to the hardware. The user can simply add code to take a picture by using **Camera.TakePicture** and assigning an event handler to the **Camera.PictureReady** event. The event handler then takes whatever action is needed with the new JPEG image, sending it to a display, writing it to the SD card, or sending it to a website. Gadgeteer has handled all the low-level code to interface the module hardware to the mainboard hardware, allowing a simple API to the peripheral, so that the application writer can concentrate on the application's functionality.

As all the low-level details of the hardware have been abstracted, and a higher level common hardware interface has been exposed, there is no longer a hardware dependency. Any manufacturer's mainboard that supports serial interface sockets can be used, and the application code remains the same. All that is required is that the application be rebuilt to accommodate the type of mainboard being used.

The Application Designer

But Gadgeteer does not stop here. As well as providing a framework to simplify connecting sensors to an embedded processor, it also supplies the tools you need to configure your application, setting up the connection of the modules and creating instances of the module objects and putting them into a skeleton application for you. Visual Studio project "templates" are provided, which will generate the "boilerplate" code for the application and use a custom GUI application designer to select the mainboard and modules from a Toolbox.

To do this, it uses an add-in to Visual Studio, a GUI-based application designer. The templates and designer are installed into Visual Studio by the install package (MSI) for the Gadgeteer framework. All mainboards and modules should supply support for this designer in their firmware install packages. When you generate a new Gadgeteer application in Visual Studio, the template will open the graphic designer. You can use it to drag-and-drop installed Gadgeteer elements from the toolbox onto the designer. Each Gadgeteer mainboard or module is represented in the GUI by a photographic image of the device, with the connection socket highlighted. We will cover the details on how to use the designer and write an application in Chapter 2; this is a taster of what is to come.

Using Application Designer with the Camera Project

Let's return to the camera module project to see how the designer works. We will demonstrate the principles of the designer and a Gadgeteer application with real code. Do not worry if you don't understand the details of the code, we will cover that in detail in later chapters. The general sequence is:

1. In the designer interface, a default mainboard is added automatically. This will be either the last type of mainboard used or the first mainboard found in the Toolbox. If this is not the mainboard you wish to use, a different one can be dragged from the Toolbox and dropped onto the designer. Drag-and-drop the Required modules onto the designer from the Toolbox.

2. The designer understands all the parameters of the mainboard sockets and the socket requirements of the modules. If you hover the mouse pointer over a module connector, the designer will highlight all the possible places you can

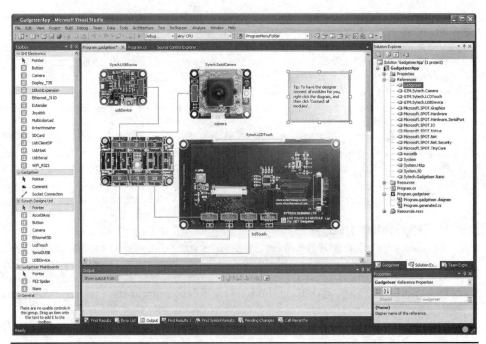

FIGURE 1-3   Graphic designer UI with possible connection places highlighted

connect to on the mainboard (see Figure 1-3). Simply drag-and-drop the connection from the module to the mainboard connector to connect. It even offers an auto-connect function that will connect up all modules for you.

3. After the modules are connected to the mainboard, the designer will generate the code stubs for your project. If you examine your project code, you will see that the designer generated code and the program class. All the modules and any required dynamic-link libraries (DLLs) have been added to the project. An instance of each module has been created, and the sockets to be used have been assigned. All you need to do now is add the code to use the functionality of the modules in your application.

4. The designer will generate the main program code file. This will be divided into two files, one with code controlled by the designer, which creates the instances of the modules and connects them to the mainboard, and one where you place your implementation. We will discuss the details of this process in the following chapters.

The following is the program-generated code, **Program.Generated.cs**. This is the application framework code created from the graphic designer project design and controlled by the graphics designer. As the warning says at the beginning of the file, do not modify this, as the designer will over write any changes you make.

```
//-------------------------------------------------------------------
// <auto-generated>
//      This code was generated by the Gadgeteer Designer.
//
//      Changes to this file may cause incorrect behavior and will be
//      lost if the code is regenerated.
// </auto-generated>
//-------------------------------------------------------------------

using Gadgeteer;
using GTM = Gadgeteer.Modules;

namespace GadgeteerApp
{
    public partial class Program : Gadgeteer.Program
    {
        // GTM.Module defintions
            Gadgeteer.Modules.Sytech.SerialCamera camera;
            Gadgeteer.Modules.Sytech.LCDTouch lcdTouch;
            Gadgeteer.Modules.Sytech.USBDevice usbDevice;

            public static void Main()
        {
                //Important to initialize the Mainboard first
            Mainboard = new Sytech.Gadgeteer.Nano();

            Program program = new Program();
            program.InitializeModules();
            program.ProgramStarted();
            program.Run(); // Starts Dispatcher
        }
```

```
private void InitializeModules()
{
        // Initialize GTM.Modules and event handlers here.
        usbDevice = new GTM.Sytech.USBDevice(1);

        camera = new GTM.Sytech.SerialCamera(2);

        lcdTouch = new GTM.Sytech.LCDTouch(10, 9, 8,
        Socket.Unused);

    }
  }
}
```

Gadgeteer creates a new Micro Framework application, and creates and initializes instances of the mainboard and the desired modules. In this example, the modules are a USB device, serial camera, and a LCD display. When you are creating the modules, the socket numbers that the modules use are passed into the constructor. We have highlighted the instances of the modules in bold text; this is for illustration only—the actual code will not do this.

Finally, the program execution is started.

We can now add application code to the **programstarted()** method in the program.cs file. The program.cs file is the part of the main application class not controlled by the designer. This is where you add your application code. We are using a feature of .NET programming called "Partial" classes. This allows a single class to be impleneted with parts in different files. The compiler will join the different files together and treat them as one in the build process. The next code sample is the implementation of the simple application in program.cs. This is where we will add application code.

In this simple example, we'll use a Gadgeteer timer to take a picture using the camera every 20 seconds. When each picture is ready, it will be displayed via the LCD display.

Because our modules have already been created by the designer, we simply need to add code to indicate how we want to use them.

1. We start by enabling the camera and then defining an event handler for the Camera Picture Ready event. This handler will send the JPEG picture to the LCD display.

2. On the display, we will use the Gadgeteer Simple Graphics functions to do all the work for us. Simple Graphics is a Gadgeteer library providing a graphics library, independent of the more complicated .NET WPF library. We will use the utility that allows an image to be drawn on a LCD display. So the call is simply **DisplayImage**, passing the image and where on the display we want it to appear.

3. Next we create the timer and set it to have a tick of 20 seconds.

4. We then implement the timer tick handler, which calls **Camera.TakePicture**, to start a picture capture process.

The Gadgeteer Framework and module firmware have handled all the low-level details for us, allowing us to concentrate on our application functionality—taking a picture every 20 seconds and displaying it on the display. All we have had to add is the code in the following code sample. As you can see, the added functionality is at a high level and we do not need to worry about any of the low-level details of how the modules work.

```
using Gadgeteer;
using Microsoft.SPOT;
using GT = Gadgeteer;
using GTM = Gadgeteer.Modules;
using Gadgeteer.Modules.Sytech;

namespace GadgeteerApp
{
    public partial class Program
    {
        //picture timer
        private Gadgeteer.Timer timer;

        void ProgramStarted()
        {
            //Enable and setup the camera
            camera.EnableCamera();
            //connect up the picture ready event handler
            camera.CameraPictureReady += camera_CameraPictureReady;

            // use a Gadgeteer timer to take a picture every 20 seconds
            timer = new Timer(20000);
            //set up tick handler
            timer.Tick += timer_Tick;

            // now start the timer
            timer.Start();

            // Do one-time tasks here
            Debug.Print("Program Started");
        }

        /// <summary>
        /// Event handler for the Camera Picture Ready
        /// Display the picture on the Display
        /// </summary>
        /// <param name="sender"></param>
        /// <param name="jpegImage"></param>
        void camera_CameraPictureReady(SerialCamera sender,
                                        Bitmap jpegImage)
        {
            // Use simplegraphics to display picture 5 pixels from the
            // left and 5 pixels from the top of the display
            lcdTouch.SimpleGraphics.DisplayImage(jpegImage,5,5);
        }

        /// <summary>
        /// Timer tick handler
        /// Fired every 20 seconds, take a picture
```

```
        /// </summary>
        /// <param name="timer"></param>
        void timer_Tick(Timer timer)
        {
            camera.TakePicture();
        }
    }
}
```

CHAPTER 2

Software Development Environment

The development environment required for writing Gadgeteer applications is based around Microsoft Visual Studio 2010 and a number of SDKs (Software Development Kits). At the time of writing, Gadgeteer hardware supports version 4.1 of the Micro Framework. The process of changing support to Micro Framework 4.2 is in progress. The install instructions will use Micro Framework 4.1 as most devices support this. The instructions equally apply to Micro Framework 4.2. The following are required for a complete development environment:

- Microsoft Visual C# 2010 Express (or Professional)
- Microsoft .NET Micro Framework 4.1 SDK
- Microsoft .NET Gadgeteer Core SDK
- Gadgeteer Mainboard and Module SDKs

The free edition Visual C# 2010 Express or better is required. The Microsoft .NET Micro Framework SDK installs the add-on development tools for Micro Framework. The Gadgeteer Core SDK extends Micro Framework for Gadgeteer development. Both these SDKs are free downloads. You will also need the Gadgeteer SDKs for any mainboards and modules you use. These will be available from the board manufacturers. Most manufacturers also maintain their mainboard and module source code in the Gadgeteer Codeplex repository.

Install Visual C# 2010 Express

Visual C# 2010 Express is available to download from www.microsoft.com/visual studio/en-us/products/2010-editions/visual-csharp-express.

1. Save the web install package vcs_web.exe to your local drive.
2. Navigate to the folder and double-click the executable to start the install process.
3. You will see a security warning dialog asking if you want to run this application, as shown next. Click Run to start the install process, which is a web-based install. You'll need a reasonably fast Internet connection.

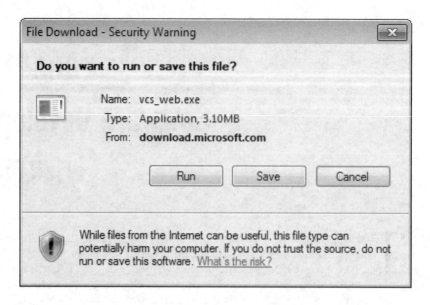

4. The install process will proceed with installing Visual C# 2010 Express from the web service. At the Welcome screen, shown here, click Next to move on.

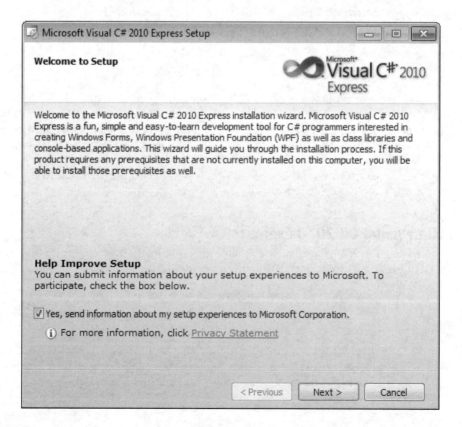

5. In the next window, the progress is displayed as Visual C# 2010 Express downloads and installs the various components, as shown here:

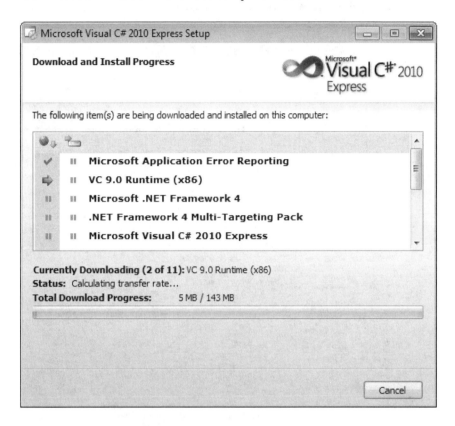

6. After the first few components have been installed, you will be requested to allow a restart of your computer. Click Restart Now, as shown. You PC will reboot and the install process will continue.

7. When the install process has installed all the required components, you will see the Setup Complete screen, shown next. The install process is quite large and the time taken to complete depends on your Internet connection speed.

8. In the Program menu, you will see a new entry, Microsoft Visual Studio 2010 Express, shown at left in the following. Right-click this and select Send to | Desktop (Create Shortcut), as shown:

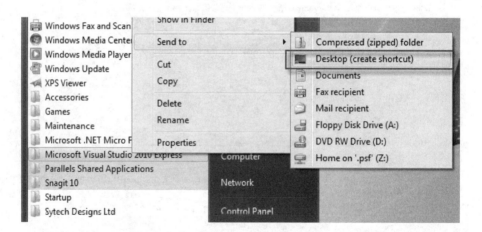

9. Select the new Visual Studio entry on the Program menu to start up Visual Studio to test the installation. Visual Studio will load and the Start page will be displayed, as shown in Figure 2-1. Then close Visual Studio and proceed with the .NET Micro Framework installation.

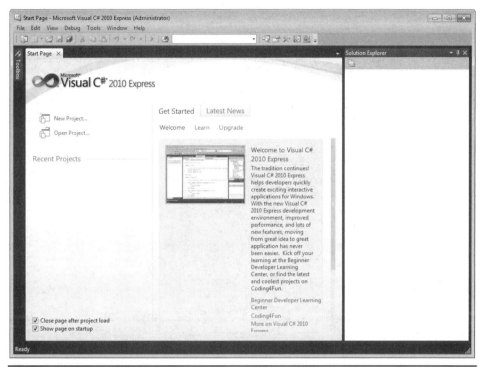

FIGURE 2-1   Visual Studio Start page

Install the .NET Micro Framework

Now let's install .NET Micro Framework.

1. Download the Micro Framework 4.1 QFE1 SDK from www.microsoft.com/
   download/en/details.aspx?id=8515, as shown here:

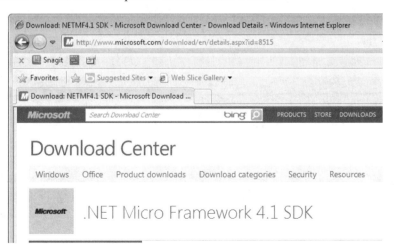

2. The SDK is a ZIP file and is about 18MB; it will be saved automatically to a folder on your PC, as shown:

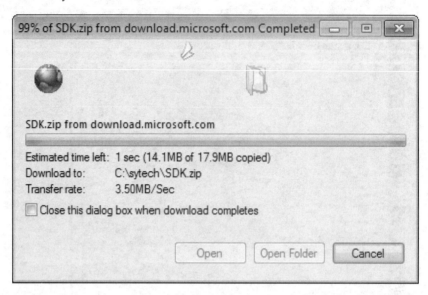

3. Unzip the downloaded SDK install file. Navigate to your unzipped folder and double-click the MicroFrameworkSDK.msi file to start the install process. Click the Run button on the Security Warning dialog.

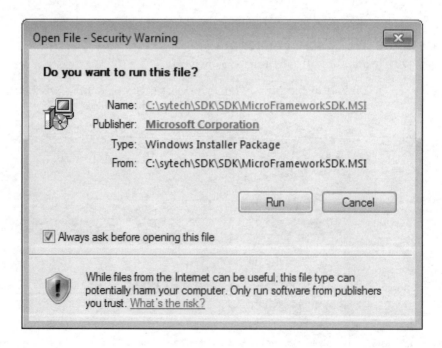

4. The first screen of the install process is shown next. Ensure it says it is the QFE1 version. Then click Next.

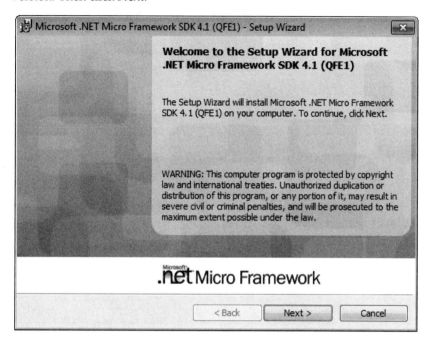

5. The installer will extract all the required components and install them, showing the progress, as shown:

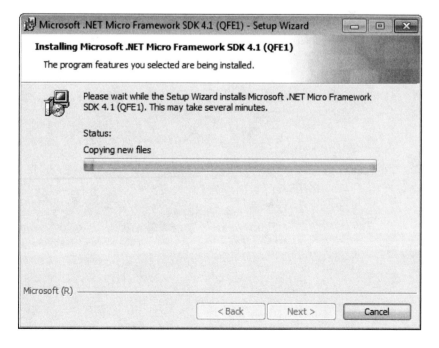

6. The final part of the installation registers all the components. This can take a minute or so. When the installation is complete you will see the following dialog telling you that the Setup Wizard has completed. Click Finish.

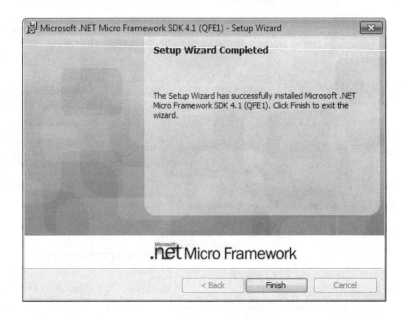

7. Start up Visual C# 2010 Express and choose File | New Project.

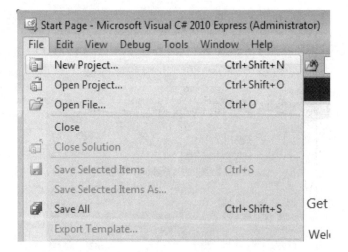

8. Under Installed Templates, you will see Visual C#. Click this to reveal a new category called Micro Framework. Select this and you will see the new project template for Micro Framework projects:

The Micro Framework SDK is now installed in Visual C# 2010 Express. Close Visual C# 2010 Express.

Install the Gadgeteer Core SDK

Start by downloading the latest Gadgeteer Core SDK from the Gadgeteer Codeplex site at http://gadgeteer.codeplex.com/releases.

1. Select the .NET Gadgeteer Core, as shown next, and download and save it to a folder on your PC. Note the version number may be a later version than shown in the screen shot—use the latest release version available.

2. Navigate to the downloaded file and double-click the msi file to install the SDK. Click the Run button on the security dialog to start the install. The install package will display the start dialog. Then click Next, as shown:

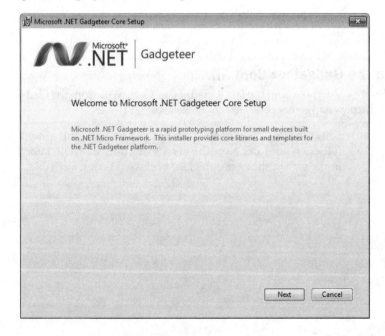

3. As the SDK is installed, a dialog will show the progress.

4. When the installation process is complete, the following screen displays, indicating that the Gadgeteer core is now installed into Visual C# 2010 Express:

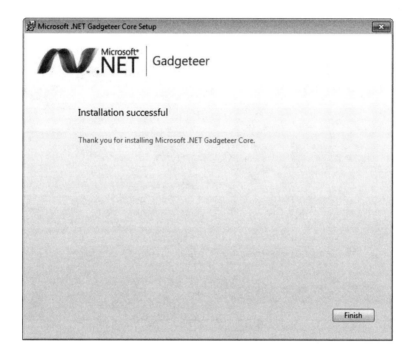

5. The Gadgeteer core is now installed into Visual Studio C# 2010 Express.

6. Open Visual C# 2010 Express and select New Project from the main menu.

7. In the Installed Templates pane, you will now see a new category, Gadgeteer, in addition to Micro Framework. Select the Gadgeteer category to reveal the new .NET Gadgeteer Application template on the right, as shown in Figure 2-2.

FIGURE 2-2   Gadgeteer Application template

Gadgeteer Documentation

It is also a good idea to install the latest Gadgeteer documentation. This is available as either an HTML-based help file or as a chm file. Both downloads can be found on the Codeplex Gadgeteer downloads page, shown in Figure 2-3.

Download your preferred help format (or both formats) into a folder on your PC. If you download the HTML reference version, unzip the file.

If you download the chm file, you will need to unblock the file before you can use it (Windows security). To do this, right-click the chm filename and select Properties to open the file's Properties dialog. In the General tab, at the lower right, click the Unblock button, as shown in Figure 2-4. The chm file is now ready to use.

The other files available for download from the Codeplex Gadgeteer site are the .NET Gadgeteer Builder Templates and the Mainboard and Module Builder Manuals.

Gadgeteer Mainboard and Module Project Templates

The .NET Gadgeteer Builder Templates are contained in an msi package that will install additional project templates for Gadgeteer hardware projects. You can use these templates to create your own custom mainboard and module firmware packages. The process of creating your own mainboards and modules is covered in Chapter 12.

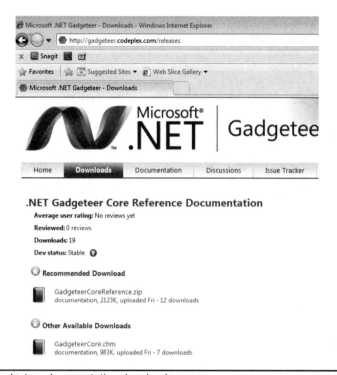

FIGURE 2-3   Gadgeteer documentation downloads

FIGURE 2-4   Unblock the chm file

The mainboard and module manuals describe the hardware and firmware rules for Gadgeteer-compatible boards.

Install Gadgeteer Mainboard and Modules SDK

The final part of your development environment is a Gadgeteer mainboard and some modules. Mainboard and module manufacturers supply an install package for their products. These can be individual install msi's for each product or a complete "kit" install package, containing several mainboard and/or module packages.

In this section, we'll install the Sytech Designs Gadgeteer Development Kit install package to demonstrate how it's done. We are using this as the example because it will just install the mainboard and module firmware. Other packages, for instance, the GHI ones, will also try and install other SDKs in addition to the modules, which can be confusing. This will add a mainboard and several modules to our design environment. You can download this install package from www.gadgeteerguy.com/downloads.

Download the install package to a folder on your PC. Navigate to the folder and double-click the msi package. This will install the mainboard and module firmware to your Gadgeteer design environment.

Tour of the Gadgeteer Development Environment

Let's take a quick tour of the Gadgeteer development tools while creating a simple project that includes a button that, when pressed, flashes an LED (solid state lamp) and writes a debug statement each time the button is pressed. This will give you a taste for the simplicity and power of the Gadgeteer design environment using Visual Studio tools. Note Visual Studio Ultimate edition is used in some of the screen shots, instead of Visual Studio Express, but all used functions are the same with the Express edition.

Starting a New Project

1. Start Visual Studio 2010 Express and select New Project.

2. In the New Project window's middle pane, select .NET Gadgeteer Application. In the Name field, replace the default "GadgeteerApp1" project name to **GadgeteerButton**. This will replace the template default project name GadgeteerApp1.

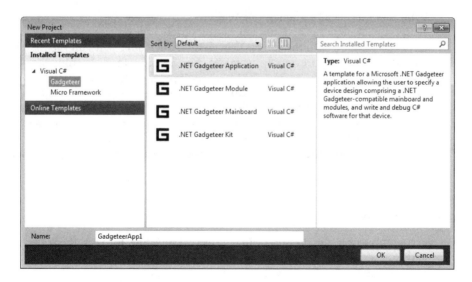

The new project will be created and the Gadgeteer designer canvas will open. You'll see a default mainboard on the designer canvas, as shown in Figure 2-5.

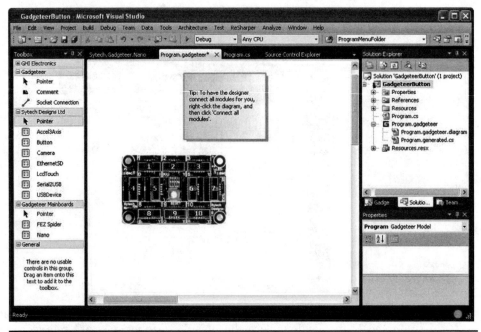

FIGURE 2-5   Gadgeteer designer canvas

Using the Canvas and Making Connections

Gadgeteer Designer is a visual interface that lets you move mainboards and modules from the Toolbox onto the canvas. The Toolbox is loaded with all the mainboards and modules you have previously installed. You select a module from the Toolbox and drag-and-drop it onto the canvas. (If you can't see the Toolbox, choose View | Toolbox.)

After you drag-and-drop modules onto the canvas, you can click the connector sockets and drag a connection from the socket on the module to a socket on the mainboard (or vice versa). When you click a socket on the module, all valid sockets for that module are highlighted on the mainboard (and vice versa).

Drag-and-drop the following modules onto the designer canvas, as shown in Figure 2-6:

- USBDevice
- Button

At this point, the designer can connect all modules for you. Right-click the canvas and choose Connect All Modules. The designer will add connections to all the modules, as shown in Figure 2-7.

To connect a module to the mainboard manually, click a module socket; valid mainboard sockets that support the module socket will be highlighted. Or click a mainboard socket, and all module sockets supported by that mainboard socket will be highlighted. To make a connection, keep the mouse button pressed and drag the connection to the socket to which you want to connect. A dotted line signifies the connection, as shown in Figure 2-8.

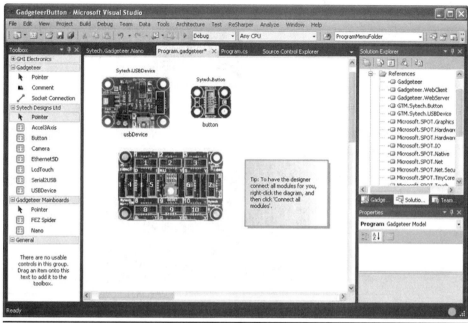

Figure 2-6   Mainboard and modules on the designer canvas

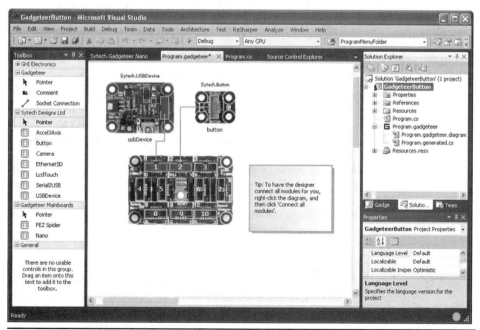

Figure 2-7   Modules connected

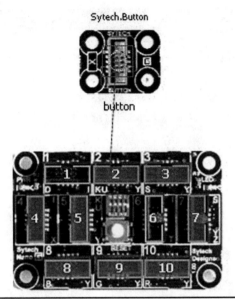

FIGURE 2-8 Socket connections

Getting Help

The Gadgeteer designer also adds a help feature tool. On the designer canvas, select a module or mainboard by left-clicking it; then press F1. The XML help documentation for that module will be displayed, as shown in Figure 2-9.

Generating Code

Using the designer, we have selected the mainboard and modules for our new project and connected them together. At this point, the designer can generate the code stubs for the new project. It will create instances of the mainboard and modules and initialize the connections. The new application will be created and the code to add the mainboard and modules will be generated, ready for the custom application implementation to be added.

The Solution Explorer window shows the new project. The designer has added the main program.cs file, the designer canvas file, and a partial class for the program.cs program. Partial classes allow code for a class to be split into different files, separating implementation into logical sections; at compile time, the separate files are effectively treated as a single file. The program.generated.cs file, shown in Figure 2-10, is auto-generated by the designer.

CAUTION *Do not make any changes to the program.generated.cs file, because the designer can overwrite your code in this file. Any code added to the program.cs file, however, cannot be changed by the designer.*

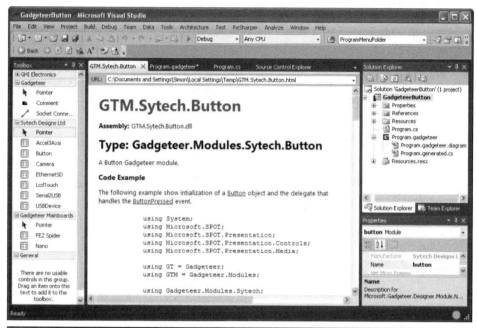

FIGURE 2-9   Help documentation

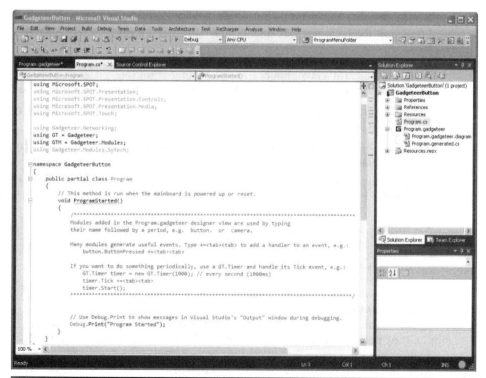

FIGURE 2-10   Auto-generated program file

Open the program-generated file, program.generated.cs, and you'll see the initialization code added by the designer. It has created instances of our mainboard and modules:

```
Public partial class Program : Gadgeteer.Program
{
    //GTM.Module definitions
    Gadgeteer.Modules.Sytech.Button button;
    Gadgeteer.Modules.Sytech.USBDevice usbDevice;
    Public static void Main()
    {
        //Important to initialize the Mainboard first
        Mainboard = new Sytech.Gadgeteer.Nano();
        Program program = new Program();
        program.InitializeModules();
        program.ProgramStarted();
        program.Run();
    }
    Private void InitializeModules()
    {
        // Initialize GTM.Modules and event handlers here.
        usbDevice = new GTM.Sytech.USBDevice(1);
        button = new GTM.Sytech.Button(2);
    }
}
```

The boilerplate code for our application has been generated by the Gadgeteer designer, and the instances of our modules have been created and connected to our mainboard:

1. The **Program** class inherits the **Gadgeteer.Program** class (supplied by the Gadgeteer Core). This means it takes and extends the **Gadgeteer.Program** class.

2. It defines the Button module as a variable **button** and the USBDevice module as **usbDevice**.

3. Then in **Main** it creates the instance of our mainboard, creates the new **Program**, and calls **InitializeModules**.

4. **InitializeModules** will create instances of our two modules, passing in the mainboard connector number they use and then assigning these new instances to our module variables (**button** and **usbDevice**).

5. **ProgramStarted** is then called, which is implemented in our Program.cs file and is where we will initialize our custom device application.

6. Then it calls **Program.Run** to start the execution of our program.

The firmware library for the Button module will create all the low-level interface code for us. It will define a GPIO pin on the processor to be an interrupt type input and connect it to the physical switch on the module. The same applies for the LED GPIO pin on the "Button" module, except this GPIO pin will be defined as an output. The Gadgeteer core knows to which socket the button module is connected, and the button module code will inform the Gadgeteer core to which pins on the socket the button and the LED are connected.

The Button module firmware will expose the high-level functions of the switch and LED to the user application. Our application will use the **Button.Pressed** event and use the module's LED functions to set the LED into a mode that automatically turns on the LED when the button is pushed. The other functions exposed by the module firmware are an event for button presses, the actual state of the button (pressed or released), and a toggle for the LED (reversing the current state of the LED to either on or off). The LED can also be directly switched on or off and its current state read. The LED can also be linked to the switch in several modes. So, in effect, we do not need to be concerned with how to turn on the LED at a low level; we just want it to turn on.

NOTE *To see the full level of support functions for the module, select the module in the designer and press F1. This will bring up the help documentation for the module's API.*

Because all the low-level connection is handled by the Gadgeteer core, only the functionality of the module is exposed to the application. Therefore, you can move the module from socket 3 to socket 4 by changing the connection in the designer, and the Gadgeteer core will change the module initialization to implement the change. So even though you are now using completely different GPIO processor pins (due to the socket change), your application code is not affected, and after a rebuild it will work as before.

Implementing the Application Code

We will now add the application implementation that uses the modules and demonstrate some of the Visual Studio 2010 development features.

For our simple example, we will set up the button module to respond to a button press event and set the LED mode to be on while the switch is pressed down. In our event handler for the button press, we are going to write a string to the debug output (of Visual Studio), saying "button pressed." The code is added to the program.cs file.

The Gadgeteer-generated code hands over control to our application at the point where the main application has been set up and the mainboard and modules have been created, connected, and initialized—at the **ProgramStarted()** method.

We need to add a handler for the button pressed event, which will be called every time the module button is pressed. We will add what we want to happen in response to this.

In the beginning of the **ProgramStarted** method, type **button** (the instance name of our button module) followed by a period (.). The Visual Studio IntelliSense feature will display a drop-down box with all the possible options for the next function or property of the button class (Figure 2-11). Select ButtonPressed. The lightning symbol indicates that this is an event.

Now type a plus sign and equal sign (like so: +=); this is a shortcut that means that the value on the left of the plus sign gets the value on the right added to it and the original value is replaced with this new sum. IntelliSense will display the possible options of items to add, which can only be an event handler method in this case. It will also suggest a name for the event handler method, as shown next. Press the TAB key to accept the suggestion.

```
*****************************************************************************/
button.ButtonPressed +=
                        new Button.ButtonEventHandler(button_ButtonPressed);    (Press TAB to insert)
    // Use Debug.Print to s
    Debug.Print("Program Started");
```

```
namespace GadgeteerButton
{
    public partial class Program
    {
        // This method is run when the mainboard is powered up or reset.
        void ProgramStarted()
        {
            /*******************************************************************************
            Modules added in the Program.gadgeteer designer view are used by typing
            their name followed by a period, e.g.  button.  or  camera.

            Many modules generate useful events. Type +=<tab><tab> to add a handler to an event, e.g.:
                button.ButtonPressed +=<tab><tab>

            If you want to do something periodically, use a GT.Timer and handle its Tick event, e.g.:
                GT.Timer timer = new GT.Timer(1000); // every second (1000ms)
                timer.Tick +=<tab><tab>
                timer.Start();
            *******************************************************************************/

            button.
```

⚡	ButtonPressed	Button.ButtonEventHandler Button.ButtonPressed
⚡	ButtonReleased	Raised when the state of Gadgeteer.Modules.Sytech.Button is low.
🔧	DebugPrintEnabled	
⚙	Equals	
⚙	GetHashCode	
⚙	GetType	
🔧	IsLedOn	
🔧	IsPressed	
🔧	LEDMode	

// Use
Debug.P

FIGURE 2-11   Button IntelliSense options

Now IntelliSense will realize that the new method does not exist, and it offers to create it for you. Once again, press TAB to accept the suggestion. IntelliSense will create the outline of the new method for you and add it to your file, as shown next:

```
*******************************************************************************
button.ButtonPressed +=new Button.ButtonEventHandler(button_ButtonPressed);

// Use Debug.Print to s    Press TAB to generate handler 'button_ButtonPressed' in this class    debugging.
Debug.Print("Program Started");
```

Now we can put the implementation in the new event handler. IntelliSense will leave a reminder, in the form of a **NotImplemented** exception, in the method. If the application tries to call this event handler, a .NET exception will be thrown to let you know there is no code in the method, rather than leaving it blank and doing nothing, so you won't wonder why nothing happens.

Now let's add a simple debug string to the event handler, which will write "Button has been pressed" to the Visual Studio Output window:

```
Void button_ButtonPressed(Button sender, Button.ButtonState state)
{
        Debug.Print("Button has been pressed");
}
```

Now we'll set the LED mode to be on while the button is pressed. Back in the **ProgramStarted** method, on the next line to the button event handler code you just added, type **button.** (including the period). IntelliSense will pop up all the button class options. Select LEDMode and an equal sign (=). When IntelliSense pops up again with the options for the LEDMode setting, select Button.LEDModes, as shown. This is an *enumeration* of all the states that the LEDModes can be in.

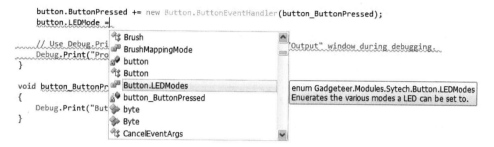

Add a . (period) and IntelliSense will pop up again, displaying the various options for the LED Modes: Off, On, and a number of dynamic mode options. These can toggle the LED when the button is pressed or released and also turn the LED on while the switch is pressed or released. Select the OnWhilePressed mode, as shown next:

```
button.ButtonPressed += new Button.ButtonEventHandler(button_ButtonPressed);
button.LEDMode =Button.LEDModes.

    // Use Debug.Print to show messa    ⬛ Off              'Output" window during debugging.
    Debug.Print("Program Started");    ⬛ On
                                       ⬛ OnWhilePressed    ┌──────────────────────────────────────────────┐
                                       ⬛ OnWhileReleased   │ Button.LEDModes LEDModes.OnWhilePressed      │
)id button_ButtonPressed(Button sen  ⬛ ToggleWhenPressed  │ The LED is on while the button is pressed.   │ state)
                                       ⬛ ToggleWhenReleased└──────────────────────────────────────────────┘
    Debug.Print("Button has been Pressed");
```

We have now added the code that will react when the button is pressed and we've set the behavior of the LED. Here's the complete code we have added:

```
public partial class Program
    {
        // This method is run when the mainboard is powered up or
        // reset.
        void ProgramStarted()
        {
                /*******************************************************
                Modules added in the Program.gadgeteer designer view
                are used by typing their name followed by a period,
                e.g. button. or camera.
                Many modules generate useful events. Type
                +=<tab><tab> to add a handler to an event, e.g.:
                button.ButtonPressed +=<tab><tab>
        ********************************************************/
            button.ButtonPressed += new
Button.ButtonEventHandler(button_ButtonPressed);
```

```
    button.LEDMode = Button.LEDModes.OnWhilePressed;
    // Use Debug.Print to show messages in Visual Studio's
    // "Output"
    // window during debugging.
    Debug.Print("Program Started");
}

void button_ButtonPressed(Button sender, Button.ButtonState
state)
{
    Debug.Print("Button has been Pressed");
}
}
```

We can now build this complete project. If you already have the physical hardware, you can deploy it. In Solution Explorer, right-click the project name and choose Build, as shown next:

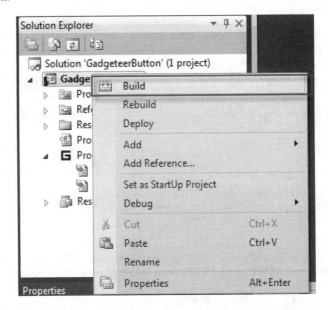

Visual Studio will now build the complete project. The progress and result of the build are shown in the Output window, as shown next. Any problems will be displayed here and in the Errors window.

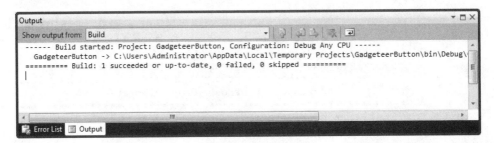

Saving Projects in Visual Studio Express Edition

The Express edition of Visual Studio handles saving your new project slightly differently than the "professional" editions of Visual Studio. In the professional editions, when you create a new project, you also have to set the save path; in the "express" edition you are not given the option to save your project when you create it. You either have to go to the main window menu options and select "Project -> Save As", or when you go to close Visual Studio, you will get a dialog box asking if you want to save the project.

Summary

That completes our quick tour of the development environment. You have seen how you can easily create a new Gadgeteer application project, and you've learned how to add mainboard and modules and how to bring up the module and mainboard online help in the graphic designer.

Finally you learned about some of Visual Studio's IntelliSense features, which help you to complete the code and prompt you with component properties.

We demonstrated the principles of a very simple Gadgeteer application, with an event-driven handler for the button press. Finally, we built our new application.

CHAPTER 3

Gadgeteer Sockets, Mainboards, and Modules

*S*ockets, *mainboards*, and *modules* are the key physical elements of a Gadgeteer application. Each of these elements is modeled in software elements that hide the physical details of the hardware and expose a common interface to the application, and this allows any modules to be used with any hardware, with a few caveats.

The most important element, the *socket*, is the core of Gadgeteer; it allows all the "magic" to happen. The socket provides the interface to the physical mainboard hardware. It holds a map between the pins on the socket and the hardware functions on the mainboard. The Gadgeteer core provides abstractions of the hardware functions associated with the socket pin, and the mainboard implements the actual interface between the Gadgeteer abstraction definitions and the actual hardware.

A mainboard, for example, must physically support analog inputs to use a module with analog inputs. The API provides a higher level general interface to the analog input functions. Different mainboards will physically interface to an analog input in different ways, but all provide the same high-level interface. To an application, the interface from processor A will be identical to that for processor B; from the application's point of view, they are the same.

Each socket on the mainboard is defined and the Gadgeteer core stores all the socket objects in an *array*. The module uses one or more sockets to interface to the physical module hardware functions. It also uses the Gadgeteer core abstraction of the supported hardware interface. So, for instance, if our module implements a magnetic compass, it would include physical hardware that uses a compass chip, and the compass chip would have an interface to a processor that is *Inter-Integrated Circuit (I2C)*–based.

NOTE *I2C is a standardized industrial two-wire serial protocol, with a clock signal and a serial data signal. It was developed by Philips (now NXP) in the early 1980s. Many sensor chips use I2C as their communications/control channel.*

The module's connection to the processor requires I2C, so a socket is provided on the module, pinned out as a Gadgeteer I2C socket (type I). The module also gets its power, via the mainboard, from the socket.

The module's physical communications/control connection to the mainboard via the I2C socket is provided by the Gadgeteer core (on top of the Micro Framework). This interface is the same, regardless of the mainboard used, as long as the mainboard implements the Gadgeteer I2C function.

A "generic" connection exists between the module's hardware and the mainboard's hardware. The module designer implements a managed driver (in C#) for the module that provides a high-level interface for the application programmer—in our compass example, this would be a function such as **ReadCompassHeading**, which would return a result in degrees. The module designer will implement the control code to read and write the required registers on the compass chip, using the I2C library provided by Gadgeteer.

Keep in mind that the module firmware does not directly use any of the hardware capabilities or API calls that are specific to the mainboard, because the firmware has no knowledge of the actual mainboard to be used. The module firmware uses only Gadgeteer and Micro Framework functions. This mechanism removes the dependency on the physical hardware, allowing the module to work with any Gadgeteer-compatible mainboard that supports I2C.

NOTE *When I refer to "sockets," I am referring to the Gadgeteer* **Socket** *class, not the .NET networking socket class, unless otherwise stated.*

Gadgeteer Sockets

In general, an application does not access a socket directly; instead, sockets are the domain of mainboard and module firmware. The application uses the API functions exposed by the module firmware.

All sockets in Gadgeteer are physically the same: a ten-pin box header, with polarized entry. Three of the ten pins are defined as power and ground, with two power pins of +5V and +3V3, and a common ground. The remaining seven pins are defined as data pins.

The data pins of a socket are configured for the physical functions they support. A socket can support multiple functions at the same time. Each function is identified by a letter. The table in Figure 3-1 shows the currently defined socket functions available and their pin usage.

A mainboard configures all its sockets as one or more of each of the supported types. A collection (array) of all the sockets on the mainboard and their supported functions are maintained by the Gadgeteer core. The mainboard will configure the pins of each of its sockets and map them to the physical hardware to achieve the supported functions for each socket.

The **Socket** class is the keystone of the Gadgeteer core but is not normally used directly by an application. Sockets are used by mainboard firmware and module firmware. A general knowledge of the **Socket** class is useful in understanding how Gadgeteer works but is not required to write Gadgeteer applications. Let's take a brief tour of the main methods and properties of the **Socket** class, which are shown in Figure 3-2.

Each socket is identified by a name and a **SocketNumber** property. The **SocketNumber** returns an integer value that is the actual socket number on the mainboard; these normally start with 1 and are sequential. The number is displayed in the box surrounding the socket on the physical mainboard. The name is the string name of the socket; this is used in any error messages generated by the functions the socket supports. The string name is normally the socket number displayed as a string.

TYPE	LETTER	PIN 1	PIN 2	PIN 3	PIN 4	PIN 5	PIN 6	PIN 7	PIN 8	PIN 9	PIN 10
3 GPIO	X	+3V3	+5V	GPIO!	GPIO	GPIO	[UN]	[UN]	[UN]	[UN]	GND
7 GPIO	Y	+3V3	+5V	GPIO!	GPIO	GPIO	GPIO	GPIO	GPIO	GPIO	GND
Analog In	A	+3V3	+5V	AIN (G!)	AIN (G)	AIN	GPIO	[UN]	[UN]	[UN]	GND
CAN	C	+3V3	+5V	GPIO!	TD(G)	RD(G)	GPIO	[UN]	[UN]	[UN]	GND
USB Device	D	+3V3	+5V	GPIO!	D-	D+	GPIO	GPIO	[UN]	[UN]	GND
Ethernet	E	+3V3	+5V	[UN]	LED1 (OPT)	LED2 (OPT)	TX D-	TX D+	RX D-	RX D+	GND
SD Card	F	+3V3	+5V	GPIO!	DAT0	DAT1	CMD	DAT2	DAT3	CLK	GND
USB Host	H	+3V3	+5V	GPIO!	D-	D+	[UN]	[UN]	[UN]	[UN]	GND
I2C	I	+3V3	+5V	GPIO!	[UN]	[UN]	GPIO	[UN]	SDA	SDL	GND
UART + Handshaking	K	+3V3	+5V	GPIO!	TX (G)	RX (G)	RTS	CTS	[UN]	[UN]	GND
Analog Out	O	+3V3	+5V	GPIO!	GPIO	AOUT	[UN]	[UN]	[UN]	[UN]	GND
PWM	P	+3V3	+5V	GPIO!	[UN]	[UN]	GPIO	PWM (G)	PWM (G)	PWM	GND
SPI	S	+3V3	+5V	GPIO!	GPIO	GPIO	CS	MOSI	MISO	SCLK	GND
Touch	T	+3V3	+5V	[UN]	YU	XL	YD	XR	[UN]	[UN]	GND
UART	U	+3V3	+5V	GPIO!	TX(G)	RX(G)	GPIO	[UN]	[UN]	[UN]	GND
LCD 1	R	+3V3	+5V	LCD R0	LCD R1	LCD R2	LCD R3	LCD R4	LCD HSYNC	LCD HSYNC	GND
LCD 2	G	+3V3	+5V	LCD G0	LCD G1	LCD G2	LCD G3	LCD G4	LCD G5	BACK LIGHT	GND
LCD 3	B	+3V3	+5V	LCD R0	LCD B1	LCD B2	LCD B3	LCD B4	LCD EN	LCD CLK	GND
Manufacturer Specific	Z	+3V3	+5V	[MS]	[MS]	[MS]	[MS]	[MS]	[MS]	[MS]	GND
DaisyLink DownStream	-	+3V3	+5V	GPIO!	GPIO	GPIO	[MS]	[MS]	[MS]	[MS]	GND

Legend
GPIO   A general purpose digital input/output pin.
(G)   In addition to another functionality, a pin that is also usable as a GPIO.
(OPT) Pin function optionally supported by a mainboard or module.
[UN]   Modules must not connect to this pin if using this socket type.
[MS]   A manufacturer specific pin, defined by the manufacturer.
!   Interrupt capable GPIO pin.

FIGURE 3-1   **Socket** types

A list of the socket's supported Gadgeteer functions can be accessed using the **SupportedTypes** property. You can read or write an array of characters; each character is the letter representing the Gadgeteer socket type—for example, an "S" represents Serial Peripheral Interface (SPI) support. You can also pass in a support type character, to a method **Socket.SupportsType**, which returns a Boolean (true or false) value to check whether a type is supported by that socket.

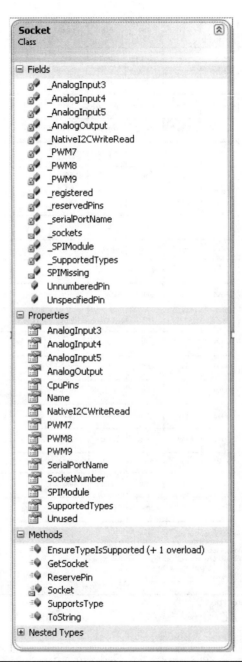

FIGURE 3-2 The **Socket** class

If a socket supports one of the serial functions (serial with or without handshake), the physical comm port name associated with the socket is stored in the socket instance. The **Socket.SerialPortName** property will return the serial port name string.

If the socket supports SPI, the Gadgeteer **SPIModule** instance (allowing access to SPI functions) can be returned using the **Socket.SPIModule** property.

If a socket has been configured to support an analog in or out function or a PWM (Pulse Width Modulation) function, the Gadgeteer interface for the function can be returned using the relevant **AnalogInput**x, **AnalogOutput**, or **PWM**x property, where the x represents the pin number for the function. Analog in is supported only on pins 3, 4, and 5, and PWM is supported only on pins 7, 8, and 9. **AnalogOutput** is supported only on pin 2.

A socket pin function can be shared among multiple sockets of the same type. SPI hardware pins, for example, will be common for any socket using the same SPI bus. An SPI clock pin could also be used as a GPIO pin on, for example, a socket that supports type "S" (SPI) and type "Y" (7 GPIO). You cannot use the same shared pin for two different functions at the same time.

If you are using the Gadgeteer designer to configure your socket connections, it won't let you assign modules to sockets that create conflicts. Suppose, for example, that two sockets use an SPI port, and one of the sockets also supports type "Y" (7 GPIO). You plug two modules into these sockets. If one of the modules tries to use the socket as a "Y" socket (7 GPIO) and the other tries to use the same socket as an SPI socket ("S"), a conflict will result. The SPI data in/out and clock pins are common to the two sockets, and one module will try and configure the socket pins as SPI pins, while the other will try and configure the pins as GPIO. The same pin can't be used for different functions at the same time.

When the Gadgeteer core is creating the module instances and initializing them in this case, it will handle one of the modules first. It will check that the required socket supports the functions required by the module, and then it will reserve the pins on that socket for use by that module. When it initializes the second module, it will go through this process again; however, when it attempts to reserve the pins, it will see that another socket has already reserved the shared pins for another use. It will then generate an exception, with a message string explaining the problem. As you are deploying the application and running it in Visual Studio to debug it, execution will stop and the exception error message will be displayed in the Visual Studio debugger output window.

In another example of two sockets, both supporting the same SPI bus functions, suppose you plug in two modules that require SPI. In this instance, the Gadgeteer core knows that the SPI pins can be shared, and it allows them to be used for the same functions.

Mainboards

The Gadgeteer core has a base class to define a mainboard, as shown in Figure 3-3. All physical mainboards are defined by a class that inherits this base class. The Gadgeteer **Mainboard** class defines a number of *abstract methods* that the physical **Mainboard** class must implement. An abstract method defines the prototype of the method but has no actual implementation of the function. The inheriting class must supply the implementation for the function.

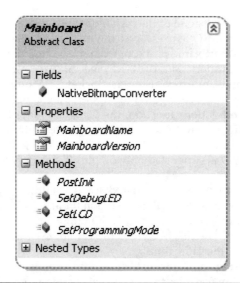

FIGURE **3-3** **Mainboard** base class

This creates an interface to a mainboard that is identical for all mainboards. The implementation of the interface will be different for each mainboard. In this way, all mainboards become the same, from an interface point of view.

In general, a Gadgeteer application does not normally access the mainboard directly. However, an application may find two properties and one function useful: the two properties are the string name and version of the mainboard, and the function is used to access the debug LED.

The Gadgeteer mainboard builder specification recommends that all mainboards include an LED for use as a debug tool; the mainboard includes a function **SetDebugLED(bool on)** for this purpose (passing in the state you wish to write to the LED as a boolean value). Calling this function with the **on** parameter set to **true** will turn the LED on; with **on** set to **false**, the LED will be turned off. The mainboard can be accessed directly using the application property **Mainboard**. The application also includes a function called **PulseDebugLED()** that will flash the LED on for 10 milliseconds. You don't really need to access the mainboard directly to flash the LED, unless you want to generate your own error or debug signals, such as three slow flashes or three fast flashes.

The remaining methods that a mainboard must implement are really for internal use by the Gadgeteer core and modules. They allow access to mainboard functions that are dependent on that particular mainboard's hardware and implementation of the Micro Framework firmware. Their main function is to allow the configuration settings of the LCD panel in use to be set. Normally, a Micro Framework device would have the LCD configuration (height, width, clock settings) set in configuration memory and they would not be changed. However, Gadgeteer has added the flexibility of being able to change the display. You may, for example, have a 4.3-inch display plugged in for one of your projects, a 3.5-inch display for another, along with a small OLED-type SPI display. The display settings are a very low-level function of the Micro Framework firmware;

memory buffers need to be configured to the correct size and graphics controller settings have to be changed. Therefore, changing the display settings will normally require a system reboot before the new settings can take effect.

Changing the settings in code will be specific for each type of mainboard hardware. The Gadgeteer **Mainboard** class defines a method called **SetLCD**, passing a set of configuration parameters. It is up to the mainboard manufacturer to implement this function in its **Mainboard** class.

Normally the display is controlled by a LCD controller in the mainboard hardware. However, an SPI or serial LCD module is built into Gadgeteer, and the graphics format may need to be changed. This involves manipulating a large data buffer, and unless this is done in native code (C/C++ at an operating system level), the function would be very slow. If the mainboard supports a native bitmap convert function, the **Mainboard** base class provides a mechanism for accessing it, via *delegation*.

Similar calls can be used to allow the debug/programming channel to be changed. The normal channel is USB, but you might want to change it to serial. A mainboard may or may not support this function. On most mainboards, this channel is set outside of the application by either switch or link settings.

The Gadgeteer core includes a custom serial interface called *DaisyLink*, a protocol designed by the Gadgeteer team. DaisyLink is discussed later in the book in Chapter 8; basically it allows modules that implement the protocol to be daisy-chained together. DaisyLink uses any mainboard type "X" socket.

DaisyLink is similar to I2C and requires a clock and serial data signal. Ideally, the mainboard should implement the protocol in native code, as it is a "bit-banged" function (that is, data transmission on a serial line is accomplished by rapidly changing a single output bit in software at the appropriate times). The base mainboard once again supplies a mechanism of delegation, so if the mainboard does have a native DaisyLink function, it can be accessed. If the mainboard does not support a native function, the Gadgeteer core will supply a managed version written in C#—however, this will be slow.

When a mainboard instance is created by the application, it will do the following:

1. Create an instance for each socket on the mainboard, assign the supported socket types, and map the processor hardware pin function to each physical socket pin.

2. Register each created socket with the Gadgeteer core.

3. Connect up the native bitmap convert function to the Gadgeteer core delegate, if the function is supported.

4. Set the name and version number properties.

Modules and Interfaces

The Gadgeteer core defines a base class for a module. This provides a common interface for the Gadgeteer core to access any module. This is part of the basic mechanism that allows the Gadgeteer core to connect any module to any mainboard. All modules must inherit from this base class, either directly or indirectly.

The Gadgeteer core also supplies three other function-specific **Module** base classes. These three module types also inherit the **Module** class. So if a specific physical module inherits one of these base classes, it also inherits the root **Module** base class. Following are the three function-specific module types:

- DaisyLink module
- Display module
- Network module

We will discuss these shortly, but first let's cover the **Module** class.

Module Base Class

The **Module** base class has a static array of modules—a global variable, if you prefer. This holds a list of all the modules the application creates. When a module is created, it is added to this list. This is mainly to ensure that something maintains a reference to each **Module** class for the life of the application.

This prevents the .NET Garbage Collector (GC) from disposing of the instances. The GC is a memory-management system. It frees the application from the responsibility of allocating and releasing memory. When an instance of a class or a variable is created, the GC will allocate the memory for it and then keep an eye on this instance or variable. When it detects that nothing is using the instance or variable anymore, GC will release the memory so it can be reused. Maintaining a list of modules for the life of the application ensures that the GC does not try and dispose of your modules.

CAUTION *Never unplug and plug in modules while the application is running (or the board is powered up).*

The **Module** class, shown in Figure 3-4, is actually very simple. It provides some logging functions and direct access (for parent classes) to the mainboard, and it ensures that the module is created on the application main thread.

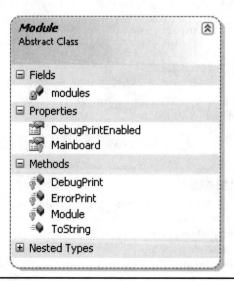

FIGURE 3-4 **Module** base class

The two module logging functions will use the debug output. Any text written to this output will appear in the Output window of Visual Studio, and while you are debugging, debug output can also be displayed in the Micro Framework PC tool MFDeploy. There are two levels of debugging output: **DebugPrint** and **ErrorPrint**. When a module uses **ErrorPrint**, the text will always be shown. **DebugPrint** can be turned on or off, using the property **DebugPrintEnabled**.

If, while developing your application using modules, you require more debugging info, you can turn on the debug print by calling **[moduleInstance].DebugPrint = true**. The content of the debug output will depend on what the module manufacturer implemented.

The mainboard property is accessible only by the inheriting **Module** class; it is not exposed publicly to the application code. However, as discussed in the mainboard section, the application code can still access the mainboard, but it doesn't really need to.

DaisyLinkModule Base Class

The **DaisyLinkModule** base class is used by modules that use DaisyLink as their connection protocol. This is a custom protocol developed by Microsoft Research for use in Gadgeteer. This module inherits the **Module** base class. It provides the protocol link for module manufacturers. If the mainboard has a native DaisyLink implementation, the module will use that; if not, a managed version is provided. The **DaisyLinkModule** base class is shown in Figure 3-5.

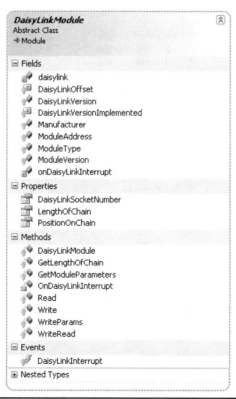

FIGURE 3-5   DaisyLinkModule base class

DisplayModule Base Class

The **DisplayModule** base class, shown in Figure 3-6, is used by graphic display modules. It inherits from the **Module** base class and adds functionality to configure the mainboard display settings configuration. It also stores the last set configuration details in nonvolatile memory. (*Nonvolatile* means the values will not be lost after a power down sequence.) It uses this setting to detect whether the current display is a new display.

This base class allows a display module to configure its required LCD settings, such as the number of pixels (picture elements or dots) in a line (width), the number or rows (height), and the clock and signal timings required by the display.

The ability to define the display requirements and automatically set them as required is a powerful feature of the Gadgeteer core and important to the concept of any module on any hardware.

The **DisplayModule** base class uses a Micro Framework feature called "extended weak references" to store the display data key parameters in a nonvolatile manner. Extended weak references allow data to be written to a special area in the mainboard's

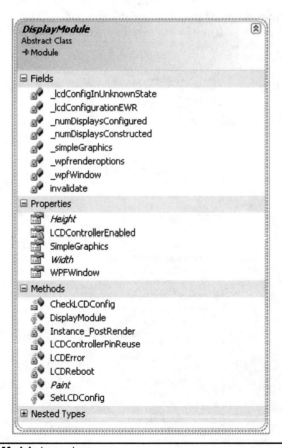

FIGURE 3-6 **DisplayModule** base class

flash (permanent memory), allowing the data to be retrieved after a power off/on sequence. However, these references are not guaranteed to be absolutely permanent; hence the "weak" part of the name. As space in this area is limited, the reference can be given a priority status—the higher this setting, the more secure the storage, if the area becomes overpopulated, some references will be removed or overwritten, starting with the lower priority ones and working up. A reference will most likely be in memory, but it's not guaranteed. The LCD setting data reference is set to the highest priority.

If this base class detects that the display settings have changed, it calls the mainboard **SetLCD** function, passing in the LCD configuration. Usually, after display settings have been changed on the mainboard, a reboot is performed.

Graphics Interface

An even more important feature of the **DisplayModule** class is the simple graphics interface, available for use by the application. In addition, the class allows the use of displays that interface with the mainboard but don't use the mainboard's LCD controller hardware. This also allows mainboards that don't support display hardware to be used with a display. Examples of this type of display are OLED SPI interface displays.

The constructor for the **DisplayModule** class allows the parameter **wpfRender options** to be passed in. This parameter setting determines whether the display rendering will use internal Micro Framework rendering (using the internal display controller of the mainboard) or will handle the rendering of the display externally (by the inheriting **DisplayModule** class). For a module to handle the rendering itself, it just adds its own implementation for the "paint" method. So a SPI-based display will take the raw bitmap data and format it as required, and then send it serially to the display using SPI. To do this, it may need to reformat the graphics format (the way the color pixel data for red, green, and blue is represented in the bitmap). You might remember from our discussion of the **Mainboard** base class that the mainboard can supply a native method to convert bitmap formats. The **DisplayModule** class (like all classes derived from module) has direct access to the current mainboard in use.

The class also allows graphics development using Windows Presentation Foundation (WPF), a stripped-down version of the desktop .NET library and an alternative simple graphics interface. The class can supply WPF windows for use by the application. The window returned will be formatted to fit the size of the display. The class can be used by displays that use the normal internal graphics controller and by displays that handle rendering themselves.

The **SimpleGraphicsInterface** class, Figure 3-7, provides a lower overhead (than WPF) graphics library, to write directly to the display. The class provides a canvas that fits the size of the display and library methods to perform graphics functions on the canvas.

Briefly, the simple graphics interface allows you to set a background color, display text, draw graphics primitives, rectangles and ellipses, set a single pixel, and draw images (bitmaps).

NetworkModule Base Class

The **NetworkModule** base class, shown in Figure 3-8, is used by modules that implement network functionality, such as Ethernet modules and WLAN modules. It adds

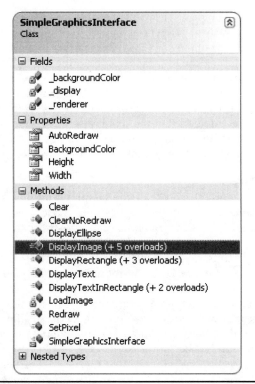

FIGURE 3-7   SimpleGraphicsInterface class

functions to get the current network settings (IP address, gateway, and so on), and set the network settings, static or DHCP IP address, mechanisms, and the status of the network.

Gadgeteer Application

Gadgeteer adds a class to encapsulate a Micro Framework application, with extensions to accommodate Gadgeteer functions. The Micro Framework "Application" class is used as the base class. This Micro Framework class provides the interface layer into the operating system, providing access to threads, the scheduler and the .NET API functions. The Gadgeteer class extends this class to add the Gadgeteer Framework functions.

Program Base Class

The Gadgeteer **Program** class implements the functionality to run your user application as a .NET Micro Framework application. It is the base class for a Gadgeteer application. This class contains a .NET Micro Framework application class and a function to start the application running. The **Program** class is shown in Figure 3-9.

A Micro Framework application runs in its own thread, and some elements of the Micro Framework library—mainly graphics operations—need to be run on this main thread. The **Program** class provides a simple mechanism that checks that an external function call is being called on the main thread; if not, it marshals (or moves) the call over to the main thread.

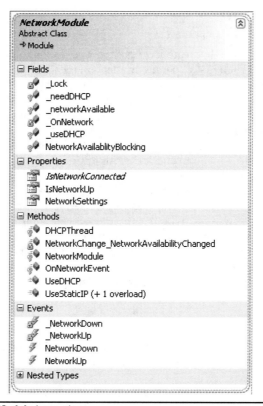

FIGURE 3-8   NetworkModule base class

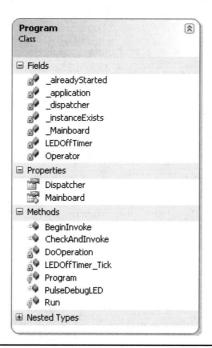

FIGURE 3-9   Program base class

It is good practice to create loop polling operations to detect some dynamic event such as a sensor change, to operate on their own background thread, preventing them for interfering with the main thread of operation. Suppose, for instance, that a module sensor change is detected on its own thread. When the user application sees this sensor change event as part of its processing, it needs to write some information to the display or change a graphic. If the handler for this sensor event is operating on a background thread, any calls to the display will not be on the main thread. The **Program** class provides a simple mechanism to check the call to update the display, and if it's not on the main thread, the mechanism will marshal it to the main thread so it can execute correctly.

This is achieved by calling the program **BeginInvoke** or **CheckAndInvoke** method. **BeginInvoke** will marshal the passed call to the main thread and execute it. **CheckAndInvoke** will test whether the passed call is on the main thread. If it is, the method returns true (but does not execute the call). If the passed call is not on the main thread, the method will marshal it onto the main thread and then execute it.

The **Program** class also contains a property with the current mainboard instance in use and a function to pulse the mainboard's debug LED.

Application

The normal (recommended) way to create your application is to use a Visual Studio Gadgeteer Application template and the Gadgeteer designer. This will create a project for your application and generate the **Program** classes, and create and initialize all your required modules.

The main class is implemented in the **Program** class in the Program.cs file. This class inherits from the **Gadgeteer.Program** base class. The mainboard and module instances will be created by the designer and placed in a partial class (code part of the program class, but just in another file) called Program.generated.cs.

NOTE *Do not modify anything in the Program.generated.cs file. The code here is auto-generated by the designer, so any changes you add will be overwritten at some point by the designer. Any code you add is in the Program.cs file.*

Here's the general program flow in the designer-generated section:

1. Create the instance of the mainboard.
2. Create the instance of this **Program** class.
3. Create and initialize all the modules, assigning them to their sockets on the mainboard.
4. Call **ProgramStarted** in the Program.cs file, where you add your application code.
5. Start the .NET application by calling **Run**.

You might be wondering how a class can create itself—a kind of chicken and egg scenario! The Micro Framework is a .NET runtime and architecturally it follows its big brother, the desktop .NET Framework.

The runtime will look for a static method (effectively a global method that exists outside of a created instance of a class) called **Main()**. This is the execution entry point of the application. This method is supplied in Program.generated.cs. Steps 1–5 are implemented in the static **Main** method.

The following code snippet is typical Program.generated content:

```
public partial class Program : Gadgeteer.Program
    {
        // GTM.Module definitions
      Gadgeteer.Modules.Sytech.SerialCamera camera;
      Gadgeteer.Modules.Sytech.LCDTouch lcdTouch;
      public static void Main()
        {
            //Important to initialize the Mainboard first
            Mainboard = new Sytech.Gadgeteer.Nano();
            // Here we create the instance of the program class
            Program program = new Program();
            program.InitializeModules();
            program.ProgramStarted();
            program.Run(); // Starts Dispatcher
        }

        private void InitializeModules()
        {
            // Initialize GTM.Modules and event handlers here.
            camera = new GTM.Sytech.SerialCamera(2);
            lcdTouch = new GTM.Sytech.LCDTouch(10, 9, 8,
            Socket.Unused);
        }
    }
```

The following code snippet is from the Program.cs file and shows the **ProgramStarted()** implementation:

```
public partial class Program
    {
        // This method is run when the mainboard is powered up or
        // reset.
        void ProgramStarted()
        {
            // This is where you add your application or entry to the
            // classes that implement your application
            // But the application must be asynchronous, or in its own
            // thread You need to exit this function to start the main
            // application running
            camera.OnPictureProgess +=
              new GTM.Sytech.Camera.PictureProgressDel(
              camera_OnPictureProgess);
            camera.CameraPictureReady +=
              new SerialCamera.CameraEventHandler(
              camera_CameraPictureReady);
            camera.DebugPrintEnabled = true;
            camera.EnableCamera();
            // Use Debug.Print to show messages in Visual Studio's
            // "Output" window during debugging.
            Debug.Print("Program Started");
        }
```

Gadgeteer Interfaces, Utilities, and Services

Gadgeteer supplies interfaces into hardware functions such as SPI, utilities such as file reading functions, and services such as the Web Server, for use by your application. We will now have a brief tour of the main interfaces, utilities, and services available.

Interfaces

The Gadgeteer core supplies a number of interfaces that encapsulate specific hardware functions such as SPI, I2C, Digital I/O, and so on. In general these are used by module implementers and provide a standardized way to access the mainboard hardware features. As an application writer, you might see one or two of these exposed in module firmware, but in general they are invisible to you.

The module implementer uses interfaces to access the lower level hardware functions, but they expose higher level functions to the application. For instance, on a Button module, the module implementer will be using the interface **DigitalInput** or **InterruptInput** to detect switch presses and pushed/released state. But they would expose higher level functions to your application, such as an **OnButtonPressed** or **IsButtonPressed** property.

Utilities

The Gadgeteer core provides a number of utilities to help in building your application.

Timer Class

The **Timer** class, shown in Figure 3-10, operates on the main thread; this means that the tick event (called when the timer period expires) will execute on the main thread. Any

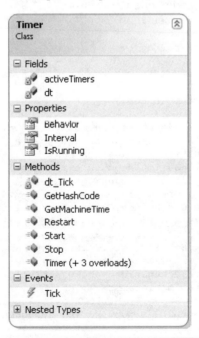

FIGURE 3-10 **Timer** class

code in your tick handler can safely access display functions and other functions without the need to be marshalled onto the main thread.

The Timer operates in two modes: **RunOnce** and **RunContinuously**. Set the Timer to the **RunOnce** mode for single timer events. Set the Timer to **RunContinuously**, the default, for cyclic period events.

To use a Timer, set the period you want the Timer to run and the run mode. Next supply an event handler for the tick event. The tick event will be called every time the Timer period expires. The Timer can be started, stopped, and restarted as required.

The **Timer** class also supplies a static method that is available even without a instance of a Timer and that can return the current machine time, from the mainboard.

In general, it is recommended that you use the Gadgeteer Timer instead of a Micro Framework Timer.

StorageDevice Class

This class, shown in Figure 3-11, is a helper class for using storage devices such as a Secure Digital (SD) card. Normally a Storage type module implementer will create the instance of the **Storage** class and allow the instance to be accessed by the application.

The class provides a number of methods to create directories, list directories and files, and read and write files.

Picture Class

The **Picture** class, shown in Figure 3-12, is a helper class for graphics handling. It encapsulates a picture or an image. Allowing access to the raw picture data as an array of bytes the encoding format of the data is also avaialable; GIF, JPEG, and bitmap are sup-

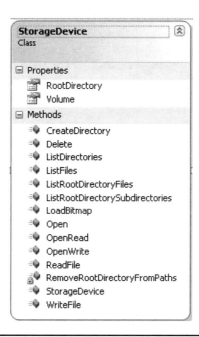

FIGURE 3-11   StorageDevice class

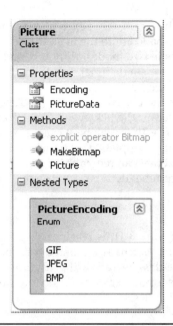

FIGURE 3-12 **Picture** class

ported. It will also convert the raw data buffer into a bitmap, irrespective of the raw data format (provided it is one of the supported encoding types).

A color helper allows colors to be accessed as a name—such as blue, green, or red, for example.

Services

The Gadgeteer core supplies some useful network services that are intended for direct use by an application:

- Web Server service
- Web Client service

These services supply and simplify access to a web server and provide client functionality. These services are discussed in more detail in Chapter 11.

CHAPTER 4

Gadgeteer API Interfaces

The Gadgeteer core provides a number of "interface" classes. These are not interfaces as in the C# language definition (a group of definitions of properties and methods provided by a class, without any implementation), but a standardized interface to access a specific hardware function. Interfaces are mainly used by module firmware writers to access hardware functions in a standardized way, irrespective of the physical hardware implementation of the mainboard. These interfaces are the highest level access methods to the hardware.

In general, they are not used by applications directly, but a general knowledge of their function is useful. Part II will offer examples of using these functions in practical projects.

Analog Input and Output

Gadgeteer defines interfaces for accessing analog voltage input and output values. These allow you to read input values and set output values. The interface API abstracts the actual hardware implementation, which will be different from processor to processor. Not all mainboards will support analog input and/or analog output functions.

AnalogInput Class

The **AnalogInput** class (Figure 4-1) encapsulates an input pin capable of reading analog voltages from 0 to 3.3V. It can return the current analog voltage value of the pin as a double with the voltage as 0–3.3V (**AnalogInput.ReadVoltage**) or as a proportion (**AnalogInput.ReadProportion**) of minimum voltage to maximum voltage of 0 to 1.

The constructor of the class configures a Gadgeteer socket pin to support the analog input interface. The socket, socket pin number, and implementing module are passed into the constructor. The socket pin used must support *analog in* functionality.

Normally, the **Socket** class creates the instance of an **AnalogInput**. When a mainboard creates a socket and assigns the pins, the socket has properties that can return **AnalogInput** instances for the relevant pins that are allowed to support this function. Of course, the underlying hardware must physically support analog in. A Gadgeteer socket can support analog input on pins 3, 4, and 5.

The accuracy and resolution of the value is totally dependent on the hardware implementation.

The function is not limited to 0 to 3.3V. If a module manufacturer implements an ADC board that reads 0–10V, the **proportion** property can be used to represent the analog input as a fraction of minimum to maximum, and then provide a function that scales this to a representation of 0–10V.

FIGURE 4-1 **AnalogInput** class

AnalogOutput Class

The **AnalogOutput** class (Figure 4-2) encapsulates an output pin capable of generating an analog output voltage. Its properties return the maximum and minimum output voltages supported. The method **AnalogOutput.Set** allows the output voltage to be set, passing in a voltage as a double. So, for example, to set the output to 2.75V, you would call *[AnalogOutput].Set((double)2.75)*, using the instance name of the class for *[AnalogOutput]*.

The class constructor passes in the socket, socket pin, and the module. Once again, it is normally the **Socket** class that constructs an instance of the **AnalogOutput** class. The mainboard pin associated with the socket pin must support analog output capability. Only one **AnalogOutput** pin is supported on a Gadgeteer socket—pin 5. However, this does not limit a module manufacturer from making a module with as many analog outputs as required; only the interface to the mainboard would be different—for example, I2C. The one analog output is a mainboard socket limitation.

FIGURE 4-2 **AnalogOutput** class

Digital Input, Output, and Input/Output

Interfaces are defined for the three gpio functions input, output, and input/output. The interfaces allow the pin operating conditions to be set and exposes high-level API functions to read and write the pins as appropriate. There is also an interface defining a gpio pin as an interrupt input function.

DigitalInput Interface

The **DigitalInput** interface (Figure 4-3) encapsulates a hardware logic input pin. The constructor associates a socket, socket pin, and optionally a module with the pin. Some hardware parameters are also set for the pin, which set whether the pin hardware uses a pull-up, pull-down, or no resistor, and if the "glitch filter" is enabled. The "glitch filter" will make the pin ignore transient changes (noise), and an input level will need to be stable at the new level for a period of time before the change is acted on. The time period for the glitch filter is dependent on the hardware.

There is one main method for **DigitalInput**: **Read()**. This will return a Boolean value that is true if the input is high (logic 1 = 3.3V) or false if it is low (logic 0 = 0V).

DigitalOutput Interface

The **DigitalOutput** interface (Figure 4-4) encapsulates a hardware logic output pin. As with **DigitalInput**, the constructor associates a socket, socket pin number, and optionally a module with the pin. It also sets the initial output state of the pin (true or false).

There are two main methods with this interface, **Write()** and **Read()**. **Write()** will set the output state of the port, using a Boolean value, to true or false. **Read()** will read the state the port has been set to (using a **Write**). It does not read an external value applied to the port; for this you need to use an input/output interface.

Input/Output: DigitalIO Interface

The **DigitalIO** interface (Figure 4-5) is a combination of the **DigitalInput** and the **DigitalOutput** interfaces. It encapsulates a hardware logic pin, which can act as either an input or an output.

The constructor associates a socket, socket pin, and optionally a module. It also sets the **Input** parameters resistor mode (pull-up, pull-down, none) and can enable the glitch filter. The initial **Output** state (true or false) is also set.

Figure 4-3   **DigitalInput** class

Figure 4-4   **DigitalOutput** class

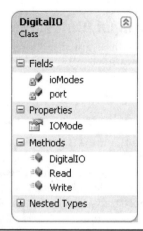

FIGURE 4-5   **DigitalIO** class

The two main methods of this interface are **Read()** and **Write()**. **Write()** behaves exactly as the output port interface and sets the state of the physical pin. **Read()** behaves exactly as the input port and reads the logic value applied externally to the port. However, you do need to inform the port which mode of operation you require. This is set with the **IOMode** property. This is an enumeration with two values: **Input** and **Output**. You can also read the property to see which mode the port is currently set to. This property controls the pin function multiplexer of the hardware.

If you need to read an externally applied value, the port mode must be set to **IOMode.Input**. If it is set to **IOMode.Output**, you will be reading the last output state set, not an externally applied input. If you need to write to the port, the value you write (true or false) will not appear on the physical pin until the mode is set to **IOMode .Output**.

InterruptInput Class

The **InterruptInput** class (Figure 4-6) encapsulates a general purpose input/output (GPIO) pin, capable of supporting interrupts. When the level changes on the input pin, the processor is notified and execution is diverted to handle the input change. This is achieved by firing a .NET event. The application supplies a handler for the event when it is triggered; the application is diverted to the handler and executes the code to handle the external event.

The constructor associates a socket, socket pin, and optionally a module to the class instance. Other parameters set are the resistor mode (pull-up, pull-down, none), whether the glitch filter is used, and the type of interrupt. The interrupt mode can be falling edge (input goes high to low), rising edge (input goes low to high), or both (triggered on either a falling edge or a rising edge).

This class extends the **DigitalInput** class, so you have access to the **Read()** method, allowing you to read the state of the input pin directly. The other main method added is **OnInterruptEvent**. This is triggered (fired whenever the interrupt mode conditions have been met—that is, the input goes from high to low) if set to falling-edge mode. This will call any delegate handlers attached to the event, passing the sender (**InterruptInput**

FIGURE 4-6   InterruptInput class

instance) and the Boolean state of the input, in the **event args** parameter. Multiple handlers can be attached to an event. They will all be executed sequentially. Event handlers are added to the event using the C# syntax +=.

PWMOutput Class

In the **PWMOutput** class (Figure 4-7), the constructor associates a socket, socket pin, and optionally a module to be associated with the pin.

FIGURE 4-7   PWMOutput class

The **PWMOutput** class encapsulates a Pulse Width Modulated (PWM) output pin. It allows either the frequency and duty cycle to be set or the period and high pulse width to be set.

The **Set()** method is used to set frequency and duty cycle. The frequency is in hertz (cycles per second) and the duty cycle is entered as a number between 0 and 100.

The **SetPulse()** method is used to set the period and the high pulse width; both are set in nanoseconds.

I2CBus Class

The **I2CBus** class (Figure 4-8) is a helper class that wraps the base Micro Framework Inter-Integrated Circuit (I2C) functionality. It allows multiple devices to be connected to the physical I2C bus and provides methods to simplify reading and writing to I2C registers on a device. Multiple external devices can be connected to the same physical I2C bus, as long as each device has a unique address. An instance encapsulates the external I2C device and attaches it to the physical I2C bus on the processor.

The constructor associates a socket and optionally a module. It also contains the I2C address of the external device and the bus frequency to use (clock rate). Multiple sockets on the mainboard can use the same physical I2C bus; the constructor will validate the use of the socket and check that the shared I2C pins are not in use as a different function (such as GPIO) on a different socket.

The main methods of the I2C bus allow a write/read operation to be performed, with the data to be written and read being passed as byte arrays. A stream read (returning a byte array) and a stream write of a byte array are also supported.

The **I2CBus** class will handle the generation of the relevant read and write transactions, and the Micro Framework will handle all the low-level protocol of handling start bits, stop bits, "acks" (acknowledgments), and "naks" (not acknowledged). Refer to the I2C protocol specification for full details.

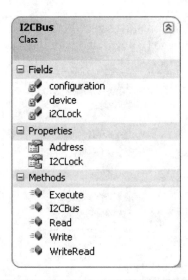

FIGURE 4-8   I2CBus class

Serial Class

The Gadgeteer **Serial** class (Figure 4-9) wraps the Micro Framework .NET serial class and adds functionality to make serial communications easier. The main feature it adds is the auto-line read capability. You can set a delimiter string of one or more characters that define the end of a receive string. A new thread will be created that monitors the received characters and stores them as a string. When the delimiter characters are received, signaling the end of the received string, a **LineReceived** event is triggered, passing the complete received string, minus the delimiter characters.

It also adds a **WriteLine** method that will terminate the string with the delimiter string and write the complete buffer to the serial port. In addition to WriteLine, several overridden versions of **Write()** are available, allowing strings and byte arrays to be written to the serial port.

The class will also manage the **OnDataReceived** event and the **LineReceived** event and marshal (move) these events to the main GUI thread. This allows the data in the event handlers to be directly used on graphics displays, and so on, as the execution is now on the main thread.

All the normal .NET Micro Framework serial class properties and methods are also available.

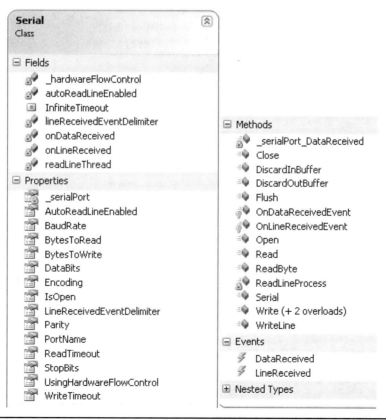

Serial
Class

□ Fields
- _hardwareFlowControl
- autoReadLineEnabled
- InfiniteTimeout
- lineReceivedEventDelimiter
- onDataReceived
- onLineReceived
- readLineThread

□ Properties
- _serialPort
- AutoReadLineEnabled
- BaudRate
- BytesToRead
- BytesToWrite
- DataBits
- Encoding
- IsOpen
- LineReceivedEventDelimiter
- Parity
- PortName
- ReadTimeout
- StopBits
- UsingHardwareFlowControl
- WriteTimeout

□ Methods
- _serialPort_DataReceived
- Close
- DiscardInBuffer
- DiscardOutBuffer
- Flush
- OnDataReceivedEvent
- OnLineReceivedEvent
- Open
- Read
- ReadByte
- ReadLineProcess
- Serial
- Write (+ 2 overloads)
- WriteLine

□ Events
- DataReceived
- LineReceived

⊞ Nested Types

FIGURE 4-9   Serial class

SPI Class

The **SPI** class (Figure 4-10) wraps the Micro Framework SPI class and adds functionality to make reading and writing to SPI devices simpler. SPI is another standard serial interface to peripheral devices. It is a synchronous serial standard and uses a serial clock, data in, and data out signals. There is also a dedicated chip select signal to enable the chip you are talking to. Multiple devices can be connected to the same clock, with

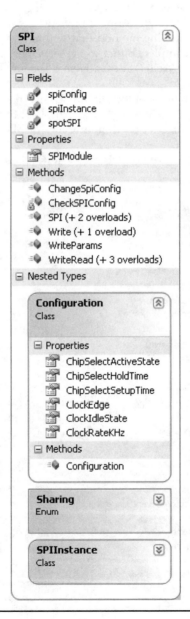

FIGURE 4-10 **SPI** class

data in and data out signals, and a different GPIO supplies the chip select to each device. The class implements a master device, which is the controller device; connected devices are slaves. The master is in control of all the communications. Devices do not initiate communications; they just respond to the master.

The class constructor creates an SPI channel to a specific device and connects it to the mainboard SPI bus. It associates a socket, the device configuration, the chip select pin to use, sharing mode, and optionally a module to the SPI bus for that particular socket. Multiple SPI buses can be provided by a mainboard. The mainboard will configure which physical bus is associated with which sockets.

The device configuration is a class with the details of the interface to the device. This contains the clock frequency to use, the timing parameters of the clock signal, and the chip select pin. This allows devices with different clock rates, clock polarities, and so on, to be used on the same SPI bus.

To communicate with the device, the **SPI** class provides methods to read a buffer of data and read/write to the device. All the low-level handling details of the bus access are handled for you, using the device configuration settings.

CHAPTER 5

Gadgeteer Mainboards and Modules

Your choice for a .NET Micro Framework–based processor board, or mainboard, depends upon the type of application you are writing and your budget. The lower cost mainboards have limited memory; some mainboards are fast with large memory; some have specific functionality such as analog and controller-area network (CAN) capability.

A large number of Gadgeteer modules are available, and new ones are regularly released. The largest single source of modules is GHI Electronics in the United States. This company is the main distributor for a number of module manufacturers as well as their own design modules. On my last check of their website, they offered more than 50 types of modules.

Gadgeteer Mainboards

Many types of mainboards are available from a variety of manufacturers. The following descriptions offer a brief overview of a few mainboard manufacturers and distributors and their products.

GHI Electronics

GHI Electronics is based in the United States and manufactures a large range of Micro Framework, Gadgeteer, and embedded products. GHI is also the main distributor of Seeed Gadgeteer modules and the Mountaineer Group Gadgeteer mainboards.

FEZ Spider

Specs: NXP LPC2478 72 MHz ARM 7 processor, 16MB RAM, 4.5MB Flash

The Spider is based on GHI's OEM Micro Framework module, the EMX. The EMX module uses GHI's commercial Micro Framework port and extends the OS firmware with many extra proprietary features and drivers, such as a USB host, which allows it to access thumb drives, mice, keyboards, and many other USB devices. It also supports the GHI WiFi module.

The device has lots of Flash and RAM memory, allowing it to handle large complex applications.

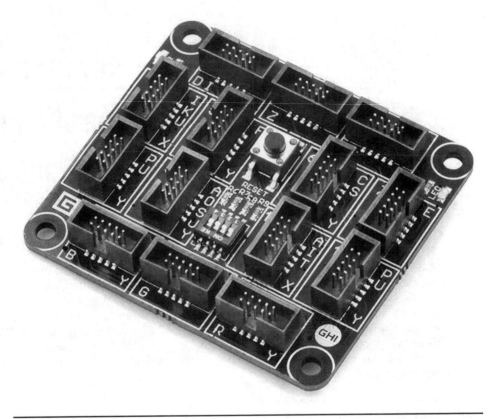

FIGURE 5-1   FEZ Spider

Shown in Figure 5-1, the Spider has 14 Gadgeteer sockets, supporting 4 serial UARTs, 2 SPIs, I2C analog in/out, PWM, USB host and device, Ethernet, CAN, SD card (4-bit interface), LCD, touchscreen, and GPIO.

FEZ Hydra
Specs: AT91SAM/RL 240 MHz ARM 9 processor, 16MB SDRAM, 4MB Flash

The Hydra was designed specifically for Gadgeteer and is completely open source. The open source firmware does not include the proprietary features of the Spider. This board uses an ARM 9 processor and has lots of Flash and RAM memory; however, the Flash memory is SPI based and not a parallel interface.

Shown in Figure 5-2, the Hydra has 14 Gadgeteer sockets supporting, 4 serial UARTs, 2 SPIs, I2C analog in/out, PWM, USB device, SD card, LCD, touchscreen, and GPIO.

Ethernet support is available using GHI's SPI interfaced Ethernet module.

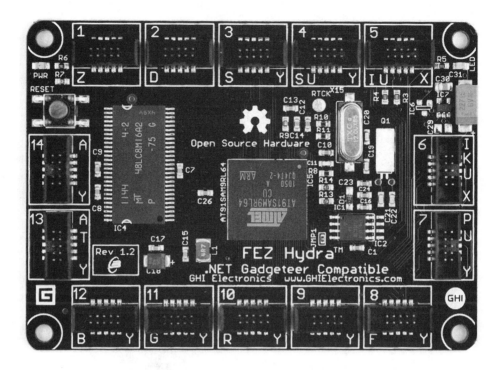

Figure 5-2   FEZ Hydra

FEZ Cerberus

Specs: STM32F405 168 MHz Cortex-M4 processor, 1 MB Flash, 192KB RAM

The Cerberus is a "system-on-chip" device. All memory is integrated into the microprocessor chip, allowing a simple, low-cost design. The onboard Flash memory is large enough to contain the OS firmware with enough memory remaining for small applications.

This is a completely open source design, with the Micro Framework port based on the Oberon Microsystems Cortex M firmware. The STM chip is the latest in the Cortex family, and its main limitation is the amount of available memory. This constrains the device to small projects with low memory requirements.

The Cerberus (Figure 5-3) has eight Gadgeteer sockets supporting two serial UARTs, SPI, I2C analog in/out, PWM, USB device, SD card, LCD, CAN, and GPIO.

Ethernet support is available using GHI's SPI interfaced Ethernet module.

Mountaineer Group

The Mountaineer Group is a partnership between two Swiss companies, Oberon Microsystems and CSA Engineering. Cuno Pfister and his team from Oberon Microsystems are behind the port for the Cortex-M3 and have contributed the sample ports for the STM M3/M4 family to the Micro Framework porting kit.

FIGURE 5-3   FEZ Cerberus

Mountaineer USB Mainboard

Specs: STM32F407 168MHz Cortex-M4 processor, 1MB + 8MB Flash, 192KB RAM

This board uses the ST Cortex-M4 processor, with an additional 8MB of external Flash memory. The board has 1MB of Flash internal to the chip, which is used for the OS and also your application code. It has an additional 8MB of Flash connected to the SPI port. This can be used for data storage—but not application code. Currently this Flash can be accessed directly using a managed code SPI driver. It is likely that in the future, a "native" filesystem will be added to access this memory. You still need to operate within the limited RAM. It is unusual in that the USB device and power supply are integrated onto the board, so no external USB device/power supply module is required. It also has a "user" pushbutton integrated on the mainboard. It is currently the smallest mainboard available (but only because it breaks some of the mainboard builder rules!).

Shown in Figure 5-4, the board has eight Gadgeteer sockets supporting four serial UARTs, two SPIs, I2C analog in/out, PWM, CAN, and GPIO. A ninth socket is manufacturer-specific and exposes the JTAG interface to the processor.

The board offers no SD support; the onboard SPI 8MB Flash is available for use for data storage at an application level. Currently there is no firmware support for CAN or a USB host.

FIGURE 5-4   Mountaineer USB mainboard

Mountaineer Ethernet Mainboard
Specs: STM32F407 168MHz Cortex-M4 processor, 1MB + 8MB Flash, 192KB RAM

This board is based on the USB version, but it adds Ethernet capability. It has USB device, power supply, and a "user" pushbutton integrated on the mainboard.

The Ethernet uses the processor's Ethernet controller and implements a fast 100 MHz Ethernet interface. This is a power board for small Ethernet-connected applications.

The board (Figure 5-5) has seven Gadgeteer sockets supporting four serial UARTs, two SPIs, I2C analog in/out, PWM, CAN, and GPIO. An eighth socket is manufacturer-specific and exposes the JTAG interface to the processor.

As with the USB version of the board, no firmware support for CAN or SD card is currently provided.

FIGURE 5-5 Mountaineer Ethernet mainboard

Love Electronics

Love Electronics is a United Kingdom–based company with a range of sensor modules and now a high-end Gadgeteer mainboard, the Argon R1.

Argon R1

Specs: NXP LPC1788 120 MHz Cortex-M3 processor, 128MB Flash, 32MB RAM

This is a full-featured mainboard that offers a huge amount of Flash and RAM.

Shown in Figure 5-6, the board has 14 Gadgeteer sockets supporting a USB device, USB host, three serial UARTs, SPI, I2C analog in, PWM, CAN, LCD, touchscreen, and GPIO. There is also a standard JTAG pin header. A special feature of this board is fast direct access to a double buffered video (display) memory. This makes the board a good choice for graphics-type applications.

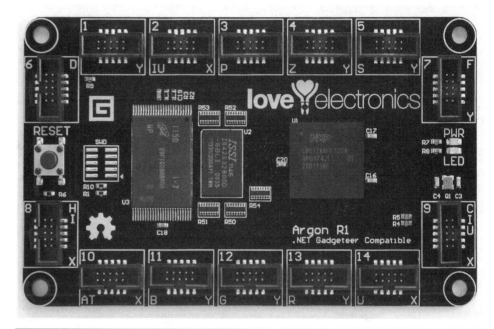

Figure 5-6   Argon R1

Sytech Designs Ltd.

Sytech Designs is based in the United Kingdom. Sytech is the designer and manufacturer of a range of OEM Micro Framework boards and Gadgeteer mainboards and modules, as well as hardware and software design consultants.

NANO Mainboard

Specs: Freescale i.MXL 200 MHz processor, 8MB Flash, 8MB RAM

The NANO mainboard is based on a Device Solutions Meridian MXS module. It has a 200 MHz ARM 9 processor with 8MB of fast Flash and RAM. This very fast and powerful Micro Framework board is capable of handling large, complex applications with high memory usage. It's also one of the smallest mainboards available (only the Mountaineer USB board is slightly smaller).

Shown in Figure 5-7, the board has ten Gadgeteer sockets supporting a USB device, two serial UARTs, SPI, I2C, LCD, touchscreen, and GPIO. One of the sockets is a manufacturer-specific socket and supports the Sytech Ethernet and SD card module. The board has default support for a 4.3-inch LCD display and touchscreen.

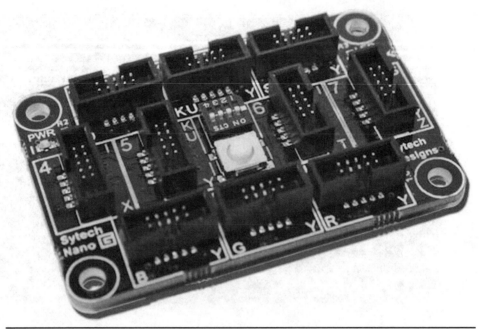

FIGURE 5-7 NANO

Gadgeteer Modules

To use a module, the mainboard must support a socket of the correct type. A module for CAN will not work on a mainboard with no CAN support, for example. The majority of modules require an SPI, I2C, or serial UART socket. All mainboards support these modules.

Ethernet, WiFi, and SD Cards

Some module functions are intrinsic to the mainboard's implementation of the driver in the Micro Framework. These are mainly SD card and Ethernet functionality. In general, you need to use the same manufacturer's module and mainboard.

Currently only the GHI Spider board supports a native WiFi module. It is possible to use WiFi on any mainboard, but with limitations. The simplest WiFi interface that will work on all mainboards is the WiFly module. This WiFi module has an integrated TCP/IP stack and a serial interface. Some versions of this module have a physical interface of a Digi XBee module PCB layout. These can be plugged into one of the Gadgeteer XBee interface modules, available from GHI and Sytech Designs.

Graphics Displays

The current mainboards that support native LCD displays are the Spider and Hydra (GHI), the NANO (Sytech), and the Argon R1 (Love Electronics).

The Spider and Hydra offer default support for the GHI 3.5-inch display, and the NANO offers default support for the Sytech 4.3-inch display. The four mainboards mentioned will work with either the GHI 3.5-inch display or the Sytech 4.3-inch display. However, the 4.3-inch display has quite a lot more pixels (picture elements) than the 3.5-inch display and works best with a faster processor, such as the one on the NANO and Argon R1.

SPI interfaced displays can be used on most mainboards. However, when using them on the limited memory (RAM) boards, their capabilities are limited. The mainboard really needs to support the required bitmap format converter routines required by these displays, to be implemented in native code. If the Gadgeteer-supplied managed (C#) routines have to be used, the display will have a slow update rate.

I2C and SPI Modules

A huge range of sensor modules use these interfaces, such as accelerometers, compasses, gyros, and so on. The problem is that the Gadgeteer standard states that modules (apart from DaisyLink modules) cannot be daisy-chained; they must plug directly into the mainboard. You will find that you will never have enough I2C or SPI sockets, no matter how many a manufacturer includes on the mainboard (both I2C and SPI are daisy-chain type protocols).

The Gadgeteer solution to this problem is to use DaisyLink modules, but the problem here is that DaisyLink is a Gadgeteer-based serial protocol that requires an active device (microprocessor) on the module to implement the protocol. This involves a lot more work to produce a module, and the DaisyLink module range is limited. As Gadgeteer develops, I am sure we will see a change allowing a solution to expand the number of I2C or SPI modules that can be connected to a single mainboard socket.

Serial Modules

The last major category of modules are serial-interfaced modules (using a UART socket). Serial-interfaced modules range from cameras, to wireless communications devices, to GPS devices. One of the more useful modules, in these days of PCs with serial RS232 ports, is the serial-to-USB virtual serial port module. This converts a mainboard serial socket into a USB virtual serial port, allowing easy serial communications to a PC.

On the wireless side, a useful module is the XBee interface module. This accepts any of the various Digi XBee ZigBee radio modules and interfaces them to the serial socket of a mainboard. Communication with these devices is in a serial protocol, based on modem-style AT commands. There are also serial-to-WiFi modules in this format, such as the WiFly modules from Roving Networks, and GSM and GPS modules are available with serial interfaces.

CHAPTER 6

Deploying and Debugging

The combination of Visual Studio and the Micro Framework operating system provides an immensely powerful environment for writing, deploying, and debugging applications in real time. In this chapter we will explore the tools that allow us to debug our applications and deploy them. The main tool for this is Visual Studio.

In addition to Visual Studio there is also a debugging tool packaged with the Micro Framework SDK. The tool is called MFDeploy.

MFDeploy is an often-overlooked Windows desktop application that can connect to a Micro Framework device and communicate with the core Micro Framework runtime. It can display information about the device; such as build version and capabilities; deploy applications; download firmware; and perform some basic debugging capabilities. It is even possible to write your own plug-ins and attach them to MFDeploy. (This is an advanced topic beyond the scope of this book.)

MFDeploy is also a useful tool for helping to solve communication problems and resetting your device back to a clean state. MFDeploy provides us with a view into the Micro Framework port on our device, for in-depth application debugging, we use Visual Studio. We can deploy applications with MFDeploy, but it is an awkward process, requiring the application to be first installed with Visual Studio, to convert it into a binary format. The binary format is then read from the device by MFDeploy and saved to disk, where it can be used to deploy to further devices. It is intended for deploying finalized applications in a production scenario. However, Visual Studio cannot deploy new operating system firmware; for this you need MFDeploy or an application supplied by the hardware manufacturer. MFDeploy is useful for a more low-level debugging tool of the hardware and Micro Framework port, whereas the Visual Studio tools target application development.

In Visual Studio, you can set breakpoints in your code and then deploy and execute the application in near real time (there is a certain amount of overhead with the debug communications to the device) on real hardware. When the breakpoint is reached, execution is halted, and you have access to all the global variables and local variables at the point of execution. You can inspect the values, change the values, and single-step through the code. You can even wind the execution point back to execute those lines again, and change variable values to see their effect.

How to debug with Visual Studio is a huge topic, so we will just cover the key features in this chapter, allowing you to experiment further for yourselves.

MFDeploy gives a view into the Micro Framework operating system. Before we investigate MFDeploy we need a general understanding of the two versions of the Micro Framework operating system. In addition to the normal operating mode, the TinyCLR, there is also a minimal version called TinyBooter. The TinyCLR is the full version that runs your managed applications, TinyBooter is a minimal Micro Framework core that supports the debug channel and file management, but does not run applications.

TinyCLR and TinyBooter

Most Micro Framework devices support two versions of the operating system: TinyCLR and TinyBooter.

All devices support TinyCLR, which contains the Common Language Runtime (CLR), allowing .NET applications to run. This is the normal operating mode of the device. Some limited memory devices substitute their own boot loader to save memory space, but, in general, if a device needs to do this, it won't have enough resources to run the Gadgeteer framework.

TinyBooter is a minimal implementation of the runtime. It does not support running applications, but it does include the core system and the debugging channel. TinyBooter can be your lifeline to recovering a badly behaving device (or, more precisely, a badly behaving application you have deployed to the device). It allows communication with the device debug channel and lets you download new firmware.

If, for instance, you download an application to the device that locks up the device and doesn't allow you to attach the debugging channel, you cannot connect to the device to erase or replace the bad application, re-flash the device, and so on. An example to consider is an application that calls a hardware reset function in the first line of the code. The device will start up and immediately reset, and then start up and reset— forever. You cannot connect the debugger channel to remedy the situation! This is where TinyBooter will get you out of trouble.

TinyBooter mode is normally selected by a hardware link of some sort—for instance, a Sytech Nano mainboard includes a dip switch, which, if closed, selects boot-up in TinyBooter mode. The FEZ Spider mainboard also uses a dip switch, and with the FEZ Hydra, you connect a button module to a particular socket and hold the button while resetting the device. Low-level debug commands can also switch the runtime mode between TinyBooter and TinyCLR, these are used by MFDeploy. TinyBooter mode will not load any application from Flash; it will load only the core Micro Framework runtime.

If you instead boot the device in TinyBooter mode, the application (in our example above) causing the resets does not get loaded, and the debug channel can be started. Then you'll be able to connect with MFDeploy. You can erase the user application from Flash, using MFDeploy—it will erase only the downloaded application, not your core firmware. Then you can put the device back into normal boot mode (TinyCLR), reboot, and the misbehaving application will no longer run, because it has been deleted. Then you can connect up with Visual Studio and normal service is resumed.

Remember that you need to be running in TinyCLR mode to deploy and debug from Visual Studio, so don't forget to remove the hardware link or you will reboot to TinyBooter!

Using MFDeploy

Access MFDeploy in the Tools section of the Microsoft Micro Framework SDK installation. Go to *<Program Files>*\Microsoft .NET Micro Framework\v4.1\Tools, and you will see MFDeploy.exe among the files. Different versions of the Micro Framework are installed in their own "version" folder. So in the path above we are using version 4.1; if you are using Micro framework version 4.2, the parent folder containing the tools is "..\v4.2\Tools". I strongly recommend you right-click it and select Send To | Desktop to create a shortcut to the application on your desktop.

Double-click on the MFDeploy.exe to start MFDeploy (or use the new desktop shortcut you have just created).

MFDeploy Main Screen

The main MFDeploy screen is shown in Figure 6-1. It has three main sections and a toolbar. The three sections are for device control, image file management, and an output window to show information text.

Device Control

The top section of the MFDeploy screen is the Device Control section as shown in Figure 6-2. It allows selecting the device to connect to. MFDeploy can detect devices on either serial, USB, or Ethernet channels. With Gadgeteer-based hardware, the usual channel is USB.

The drop-down box on the left allows selecting the communications channel from a list (Serial, USB, or Network).

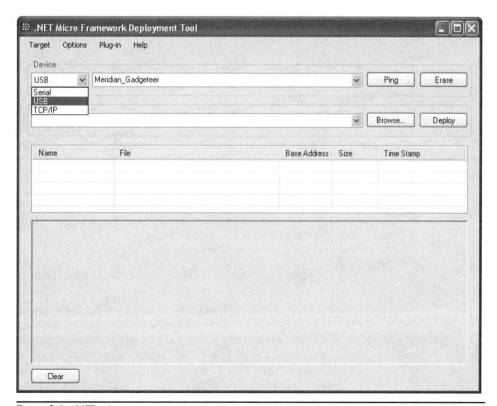

FIGURE 6-1 MFDeploy

FIGURE 6-2   MFDeploy device control section

A Friendly Name

Each Gadgeeter mainboard is assigned a USB "friendly" name, by the manufacturer, to identify it. By convention this name will be [name]_Gadgeteer, to identify it as a Gadgeteer type device. The Gadgeteer project templates set the Visual Studio project to look for a USB device with this name format, and automatically connect to it for debugging and deploying from the project. The middle drop-down list includes all the devices detected on the communications channel. The devices will be identified by their USB "friendly" name.

Connect your physical device to your PC, using USB. Select the debug channel from the Device drop-down list. Then choose the desired communication channel used to connect your device to the PC (in our example this is USB). MFDeploy will detect all devices connected using that channel communication type and display them in a drop-down list. In our example there is only one device and its USB "friendly" name is Meridian_Gadgeteer. It may take a few seconds after you first connect your Gadgeteer mainboard to the PC for Windows to detect the device and enable the USB connection. When the device is available, its name will appear in the second drop-down list. At this point MFDeploy is connected to your device.

The two buttons to the right are Ping and Erase. Click the Ping button to ping the device. The device will respond and the result will be displayed in the output area at the bottom of the window. It will show the current runtime operating mode of the device, TinyCLR, or TinyBooter. Ping establishes that the Micro Framework runtime is running on your device and that the device is responding.

The Erase button is not as scary as it might seem. It will only erase any managed application you have deployed to the device—it will not erase the Micro Framework firmware! This function is useful for putting your device back into a "clean" state, with no application loaded, just the operating system running.

Image File Management

The next section of the MFDeploy window is the Image File management area, as shown in Figure 6-3.

MFDeploy can deploy files to your device. This can be an application or new Micro Framework firmware. The files deployed are binary files, formatted for ARM processors. We will cover deploying an application from MFDeploy later in this chapter in the section "Deploy an Application Using MFDeploy." The Image File section allows you to browse and select the files you need to deploy. They are then displayed in the File list in the middle section. After clicking to select the checkbox next to the appropriate filenames, click the Deploy button to download all the selected files to the device. Figure 6-3 shows the files required to update the Micro Framework firmware:

Image File				
E:\SytechDev\MicroFramework\MeridianE\GadgeteerBuilds\mxl\1.02\NANO_MXL_4.1.40857.896 ▼		Browse...		Deploy

Name	File	Base Address	Size	Time Stamp
☑ ER_FLASH	E:\SytechDev\MicroFramework\MeridianE\Gad...	0x10020000	0x000ffb78	10/11/2011 10:35:39
☑ ER_CONFIG	E:\SytechDev\MicroFramework\MeridianE\Gad...	0x103f0000	0x000009...	10/11/2011 10:35:39
☑ ER_DAT	E:\SytechDev\MicroFramework\MeridianE\Gad...	0x10120000	0x000486...	10/11/2011 10:35:39

Figure 6-3 MFDeploy Image File section

ER_FLASH, ER_CONFIG, and ER_DAT. Because most hardware manufacturers supply a simpler, more friendly method for updating your device firmware, you usually don't need to use MFDeploy for firmware deployment.

Debug Output Area

The bottom section of the MFDeploy screen is the output area. Any text from MFDeploy or from the device is displayed here.

MFDeploy Functions

We will now look at some of the key debug functions available in MFDeploy. Many of these functions are low-level views into the low-level operating system structure of the firmware port on a device, and not really relevant to application debugging. We will just cover some functions useful for application development and debugging, such as viewing application debug statements, without using Visual Studio, checking the firmware version installed and deploying applications independently of Visual Studio.

View Application Debug Text in MFDeploy

Any debug text you add to your application—using **Debug.Print("Debug text")**—will appear in the output area of MFDeploy. (If you are connected and debugging from Visual Studio, the text will appear in the Visual Studio output window.) Let's set the debug channel.

1. Connect your device to the debug PC—normally this is with USB.
2. Start MFDeploy.
3. Set the debug channel to USB and select your device from the drop-down list.
4. From the Target menu, choose the Connect option, as shown in the following illustration.

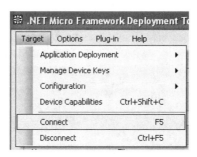

The response text in the output area will be "Connecting to [Device Name] … Connected." The debug channel is now connected to the "console" window of MFDeploy. Any debug text will appear in the window.

MFDeploy is not perfect, and sometimes you may need to kick-start the debug output. If you are not seeing the debug text, from the Target menu, choose Device Capabilities. After the capability information is displayed, you will then see your debug text. Then you can start streaming.

CAUTION *Be careful when using MFDeploy, and then using Visual Studio. If MFDeploy makes the connection to the device, Visual Studio may not always see the device. Always remember to choose Target | Disconnect to release the USB channel and allow Visual Studio to see the device.*

If You Get "Stuck" in TinyBooter Mode

Some MFDeploy actions will reboot the device into TinyBooter mode and may not reset this. A flag in the Flash configuration section will force boot-up to TinyBooter, and it is possible for this flag to get set and remain set, so that you boot into TinyBooter by default (even if your hardware is set to boot to TinyCLR). Some devices will revert to TinyCLR after a time-out period, but this can take two minutes. If, after using MFDeploy, you are finding that Visual Studio is reporting a "cannot find device" error on deploy, you may be stuck in TinyBooter mode. To check this, connect MFDeploy and do a ping. If the result is "Pinging…. TinyBooter," you know you are stuck in TinyBooter. The response to a ping will say TinyCLR if all is well.

If the boot loader flag has been set, you can use MFDeploy to reset it: From the menu at the top of MFDeploy, select Plug In | Debug | Clear BootLoader Flag. Do a test ping to confirm the change; the response should show that TinyCLR is set.

Show Device Information

You can also use MFDeploy to show information about the Micro Framework firmware loaded on your device. You can see the firmware version and get a list of hardware capabilities, such as the display settings of the LCD controller.

From the MFDeploy menu, select Target | Device Capabilities. A list of properties of your connected device will be displayed in the Console Output window, as shown here:

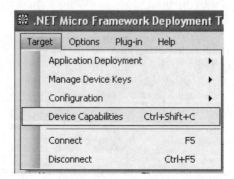

```
ClrInfo.clrVendorInfo:                    NANO_MXL_G Device
Solutions Ltd
ClrInfo.targetFrameworkVersion:           4.1.2821.0
SolutionReleaseInfo.solutionVersion:      4.1.40912.54877
SolutionReleaseInfo.solutionVendorInfo:   NANO_MXL_G Device
Solutions Ltd
SoftwareVersion.BuildDate:                Jan  5 2012
SoftwareVersion.CompilerVersion:          310836
FloatingPoint:                            True
SourceLevelDebugging:                     True
ThreadCreateEx:                           True
LCD.Width:                                320
LCD.Height:                               240
LCD.BitsPerPixel:                         16
AppDomains:                               True
```

FIGURE 6-4   Device capabilities output

A typical output window for a Sytech Nano device's capabilities is shown in Figure 6-4.

You can also get a list of DLL libraries that are loaded in Flash for your operating system build and also any DLL and application files you have currently deployed to your device. To see this information, select Plug-in | Debug | Show Device Info:

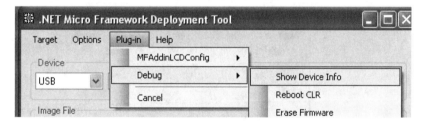

Figure 6-5 shows a typical output for device information.

```
System.Http,4.1.2821.0
Microsoft.SPOT.IO,4.1.2821.0
System.IO,4.1.2821.0
MFDpwsExtensions,4.1.2821.0
MFWsStack,4.1.2821.0
MFDpwsDevice,4.1.2821.0
MFDpwsClient,4.1.2821.0
Microsoft.SPOT.Time,4.1.2821.0
MeridianLcd,1.0.0.0
Gadgeteer,2.41.0.0
Gadgeteer.WebClient,2.41.0.0
Gadgeteer.WebServer,2.41.0.0
GadgeteerAppl,1.0.0.0
Sytech.Gadgeteer.Nano,1.0.5.0
```

FIGURE 6-5   MFDeploy device information output

Configure the Device Network Settings

MFDeploy can also be used to update the network configuration properties of your device. This allows you to set the configuration of your network adapter and WLAN settings. Depending on your hardware, these settings will be used by any Gadgeteer Ethernet and WLAN modules you use; however, these settings are hardware dependent. They can also be changed at an application level.

Choose Target | Configuration | Network, as shown:

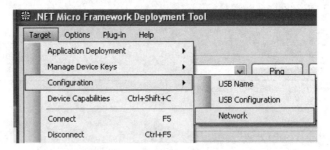

This may take a few seconds to execute, as the MFDeploy tool needs to reboot your device to TinyBooter to access the Flash configuration. When this has been done, the Network Configuration dialog box is displayed (Figure 6-6), allowing you to view all the current network settings and edit and save them back to Flash.

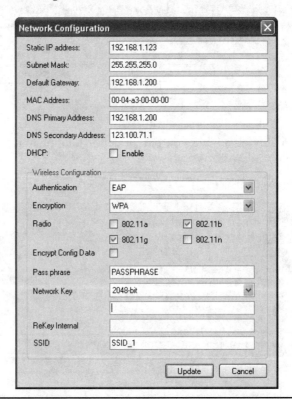

FIGURE 6-6 MFDeploy Network Configuration dialog

Deploy an Application Using MFDeploy

To deploy an application using MFDeploy, you need the file that contains your application. However, the file required is a *hex file* (binary code)—so it's not a file you've created in a Visual Studio project. You'll need to generate the hex file.

To do this, deploy the application from Visual Studio as normal. When the application is successfully deployed to the device, open MFDeploy and connect to the device. Then you can use MFDeploy to read the hex file for the application from the device and save it to the PC. This binary file can then be used as a master file to deploy the application to other devices.

NOTE *These are binary files and load the code to specific addresses with an entry point address. Use the files only on the same type of mainboard, with the same Micro Framework firmware build installed. A different type of device may well have a different memory map, and you may end up deploying the application in the wrong place and corrupting the Flash.*

To deploy the file, you need a master device with the application deployed to it using Visual Studio.

1. Connect the device to MFDeploy and select Target | Application Deployment | Create Application Deployment, as shown next:

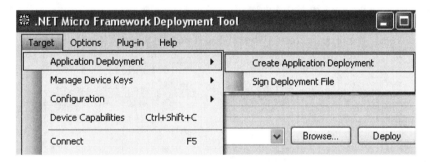

2. In the Create Application Deployment dialog box, press the browse button (labeled"…"). This will open a standard Windows file dialog window. Browse to the location where you wish to save your new file and enter a filename in the dialog.

3. We will not be signing the file, so click the OK button.

4. A progress bar will indicate your progress, as shown next. Note that the progress bar will change to 98% very quickly—but the final 2% can take a couple of minutes, so be patient—it has not crashed!

5. We have now created our binary file for the application.

6. To deploy the application to a new device, connect the target device to MFDeploy.

7. When the new device is attached, use the Browse button in the Image File area to browse to the file you created. Select the file in the window. Click the Deploy button. A progress dialog box will appear, as shown next. The first step is to connect to the device, and this can take a moment or two, because it has to reboot the device to TinyBooter. After this, the application will be deployed and then the device will be rebooted back to TinyCLR. You have just deployed your application.

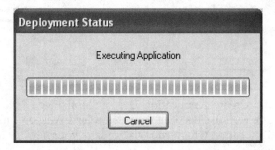

This completes our quick overview of MFDeploy. We have not covered all the features of the program, but you've seen the key elements that are useful in debugging and deploying a device.

Deploying and Debugging with Visual Studio

The first thing we need is an application to debug. We will create a simple application that uses a button module to allow a button input and a LED output.

NOTE *We briefly covered the process of creating a new project in Chapter 2, so we will not go into too much detail about creating our test project.*

1. Open Visual Studio.

2. Select New Project, select the Gadgeteer template area, and the .NET Gadgeteer Application template. This will create the new project and open up the Graphic Designer page.

3. Add a Nano mainboard (any mainboard will do) and a Button Module.

4. Use the designer to connect the Button Module to the mainboard (any valid socket will do).

5. Now we'll add our test application in the program.cs file. We will add code to respond to a button press, and then the handler will output some debug text and toggle the LED so it turns on or off every time we press the button. Our test code looks like this:

```
public partial class Program
    {
        // This method is run when the mainboard is powered up
        // or reset.
        void ProgramStarted()
        {
            // set up the button press handler
            button.ButtonPressed += new
    Button.ButtonEventHandler(button_ButtonPressed);
            Debug.Print("Program Started");
        }
        /// <summary>
        /// Button press handler
        /// </summary>
        void button_ButtonPressed(Button sender,
        Button.ButtonState state)
        {
            Debug.Print("Button pressed");

            button.ToggleLED();

            Debug.Print("Led is turned on :" + button.IsLedOn);
        }
    }
```

We added a handler method to respond to a button push. When the button is pressed, our handler method is called. It outputs some debug text: "Button pressed," in the Visual Studio output window. It will then toggle the state of the LED; if it was on it will be turned off, and if it was off it will be turned on. We then send another line of debug text to the output window, indicating whether the LED is on or off.

Build the Project

Now let's build the project.

1. In the Project Explorer window, right-click the project name and select Build. The result of the build will be shown in the Visual Studio output window, shown next, and should be successful.

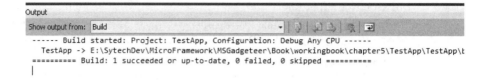

```
Output
Show output from: Build                                               ▼ | 🗐 | 🕹 🗐 | 🕱 🔁
------ Build started: Project: TestApp, Configuration: Debug Any CPU ------
  TestApp -> E:\SytechDev\MicroFramework\MSGadgeteer\Book\workingbook\chapter5\TestApp\TestApp\b
========== Build: 1 succeeded or up-to-date, 0 failed, 0 skipped ==========
|
```

2. Connect up your test device to the PC and do one of three things: right-click the project in the Project Explorer and select Debug | Start New Instance; press F5; or set the target project as the default project, right-click the project, and select Set As Startup Project. This will deploy the application to the device and start the debugger.

   Pressing the green Play arrow will also deploy and debug the main project in the Solution Explorer (the one in bold text), if you have *only one project* in the solution.

You will see a whole stream of information in the Visual Studio Output window during this process. You will see the device being rebooted, the application deploying to the device, and then a list of information about every DLL deployed. At the end of this, you will see each DLL being loaded by the debugger, and then the application will start.

At the start of the application, the Gadgeteer core will output a line of text to the output window with the name and version of the mainboard being used. You'll see a line of debug text at the start of the application, saying "Program Started." The application is now running and waiting for a button press.

Pressing the button will toggle the LED and write the debug text from the button handler. Figure 6-7 shows the debug output area after a few button presses.

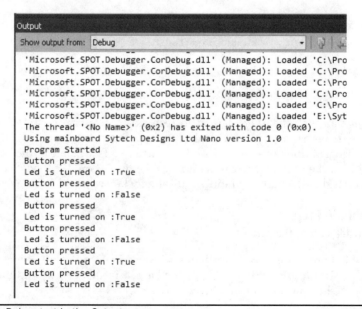

```
Output
Show output from: Debug                                    ▼ | 🗐 | 🗐
'Microsoft.SPOT.Debugger.CorDebug.dll' (Managed): Loaded 'C:\Pro
'Microsoft.SPOT.Debugger.CorDebug.dll' (Managed): Loaded 'C:\Pro
'Microsoft.SPOT.Debugger.CorDebug.dll' (Managed): Loaded 'C:\Pro
'Microsoft.SPOT.Debugger.CorDebug.dll' (Managed): Loaded 'C:\Pro
'Microsoft.SPOT.Debugger.CorDebug.dll' (Managed): Loaded 'C:\Pro
'Microsoft.SPOT.Debugger.CorDebug.dll' (Managed): Loaded 'E:\Syt
The thread '<No Name>' (0x2) has exited with code 0 (0x0).
Using mainboard Sytech Designs Ltd Nano version 1.0
Program Started
Button pressed
Led is turned on :True
Button pressed
Led is turned on :False
Button pressed
Led is turned on :True
Button pressed
Led is turned on :False
Button pressed
Led is turned on :True
Button pressed
Led is turned on :False
```

FIGURE 6-7 Debug text in the Output area

You can stop the Visual Studio debugger by pressing the blue Stop button, selecting Debug | Stop Debugging, or by pressing SHIFT-F5.

This demonstrates the most basic debug tool in our Toolbox—output text—but we have even more powerful tools to explore.

Set a Breakpoint

Now let's set a breakpoint.

1. In Visual Studio, on the program.cs page, locate the first line of code in the button press handler; it reads **Debug.Print("Button pressed")**. Left-click in the left margin next to this line; a maroon circle will appear in the margin, and the code line will be highlighted in a maroon box, as shown next. This is a breakpoint. When the code execution hits this breakpoint, it will stop and hand over the work to the debugger.

```
/// <summary>
/// Button press handler
/// </summary>
void button_ButtonPressed(Button sender, Button.ButtonState state)
{
    Debug.Print("Button pressed");

    button.ToggleLED();

    Debug.Print("Led is turned on :" + button.IsLedOn);
```

2. Restart the debugger (press F5). The application will be deployed and will start running under the debugger, as before. When the debugger hits the breakpoint, it will pause the application. The line of code to be executed next (where you placed the breakpoint) will be marked in yellow highlight, and a marker arrow will appear in the left margin:

```
/// Button press handler
/// </summary>
void button_ButtonPressed(Button sender, Button.ButtonState state)
{
    Debug.Print("Button pressed");

    button.ToggleLED();

    Debug.Print("Led is turned on :" + button.IsLedOn);
    }
}
```

We can now examine variables in real time. The button is an object, and we can now inspect all of its properties.

3. Hover the cursor over the word **button** in the line of code under the yellow-highlighted line. A drop-down box will appear, showing button {button} with a plus (+) symbol. Click the plus symbol to open the properties list.

4. You can now see all the current properties of the button object, as shown next. One property shows the status of the LED, **IsLedOn**. In the illustration, you can see that at the moment the LED is on at 0x00000001. This is a Boolean value,

which is either true or false; true values are represented as non-zero numbers and false shows a zero (0).

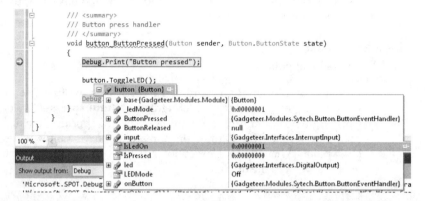

You can also open a variable or class instance in a more convenient window to inspect it and even change its setting. This is called a *QuickWatch*. Right-click the word **button** and select Quickwatch from the context menu. You'll see a QuickWatch dialog for the button, as shown in Figure 6-8. Press the plus sign, and the button properties will display. You can even select a property, and if it is writable, change its value at this point. QuickWatch is a temporary function, but you can add a permanent watch by pressing the Add Watch button or by right-clicking the variable and selecting AddWatch.

Immediate Execution

Watches allow you to look at values and even change them—which is very useful. But you can even execute methods in the debugger. For instance, at the moment, the LED is in the on state. A call to **button.ToggleLed()** will change it to off.

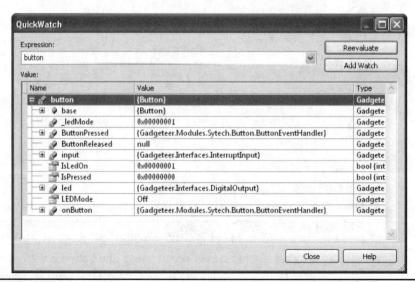

FIGURE 6-8   QuickWatch dialog

The Immediate Execution window allows you to run code then and there (while the debugger is halted or paused). You will see the Immediate Window in the bottom section of the Visual Studio screen. If it is not visible, choose Debug | Windows | Immediate (or press CTRL-ALT-I).

In the Immediate Window (we are still waiting at our breakpoint), type **button. ToggleLED();** and then press ENTER (as shown next). (Notice that IntelliSense autocompletion feature works in the Immediate Window.) You have just directly called a method on the **button** instance—the one that will toggle the LED.

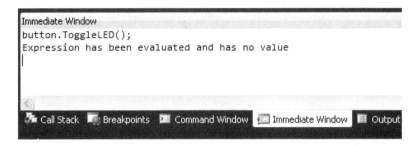

If you now go back and inspect the **button** instance, look at the property **IsLedOn** and you will see that it is now off—and if you look at the hardware, you will see that the LED has been turned off.

This ability to stop the execution at a particular point, inspect and modify variable values, and directly execute new code—even though code execution has stopped—is extremely powerful.

Single-Step Through Code and Move the Execution Point

The application code is currently stopped at the breakpoint. We can now step to the next line of code, or, if it is a call to another method, step into the next line. To do this, you can use the step into or step over symbols from the debugger tool bar, or press F11 to step into or F10 to step over. Use the step over function to step over the next two lines of code, so you end up on the final **Debug.Print** line of the button handler. As you step over, the code is actually executing—notice that stepping over the **button.ToggleLED()** line has actually changed the state of the LED.

The next line to be executed is marked by yellow highlighting, with a yellow arrow in the margin. Position the mouse cursor over this marker arrow and left-click it. Now you can move the execution point forward or backward. You can move the execution point backward (up) and actually execute the previous line of code to toggle the LED again. You can also use this method to skip a section of code, moving the execution point past it without actually executing it. You can also step over a conditional section of code, to "force your way into" the different conditions and single-step the code to test it.

More Visual Studio Features

We have just taken a brief tour of some of the key features of the debugger, but there are many more features we have not looked at. Refer to the Visual Studio debugger help documentation for descriptions of the other features, including conditional breakpoints, where you can set conditions that must also be true for a breakpoint to be executed. Another feature is the Stack Trace window, which shows you the current program flow and which method has called what to get to your current point. A Threading window shows the active thread and the other threads active in the application, and even lets you freeze particular threads (stop them from executing) so you can debug the target thread without interruption.

Visual Studio debugging is more of a hands-on topic, and the best way to learn is to run an application and experiment with the debug features.

PART II

Projects

CHAPTER **7**

Writing Gadgeteer Applications

Before moving on to specific Gadgeteer projects, this chapter will show you how to write applications for embedded projects in general, and how to write applications for Gadgeteer projects. First we'll look at the general programming models; then you'll see how you can apply these to Gadgeteer projects.

Sequential and Event-Driven Applications

Two basic application flow types are used for building applications: *sequential* (synchronous) and *event-driven* (asynchronous).

Here's an example of how a synchronous (sequential) program works:

1. Do function 1.

2. Do function 2.

3. Do function 3.

This process could even be repeated forever (Func1, Func2, Func3, Func1, Func2…).

Each function is dependent on the preceding function to finish before it executes; then the work is handed over to the next function. Each function has a start point and an end point, even if that sequence is repeated.

Here's an example of how an asynchronous (event-driven) program works:

1. If action 1 happens, do function 1.

2. If action 2 happens, do function 2.

3. If action 3 happens, do function 3.

There are no start and end points here; when a function executes is totally dependent on some action being performed, such as a button press or a sensor trigger.

An asynchronous or event-driven approach should be used to implement Gadgeteer applications. The program will be responding to sensor events, such as received data, switch closures, temperature changes, and so on, and doing something as a result of these events.

Event-driven applications can initially seem to be more complicated than sequential programs, but the latter are better solutions to an embedded control application. This becomes more critical as the application requirements get more complicated or need to be extended in the future.

Consider, for example, a GPS position logging application. Your application is required to track the location of a shipping container on its journey from point A to point B. The current position is determined and forwarded to a web service, where the current location and other information can be displayed on a map. As this is a battery-powered device, it needs to be changed to a low power mode when the container is stationary, but it needs to be able to detect when the container starts moving again. A GSM (Global System for Mobile Communications) module is required to connect to the web service over the Internet, transfer the current position, calculate the location using a GPS module, and connect to an accelerometer to detect movement.

Let's follow our application through the development process, adding functionality using both programming patterns. We'll add basic functionality, and then, over several iterations, refine and add functionality to the application. We are looking to compare developing the application with a sequential design and an event-driven design.

Sequential vs. Event-Driven Design Basics

The program loop for a sequential development process is as follows:

1. Poll the GPS receiver for the current position.
2. Calculate the container's position change.
3. If its position has changed, send data to web service.
4. Go back to step 1.

If we design this application as an event-driven one, it looks like this:

1. On receiving GPS data, the application calculates the container's position change.
2. If its position has changed, it generates a **positionChanged** event.
3. On a **positionChanged** event, if the position change is greater than x meters, the application sends the new position to the web service.

Adding Battery Power Savings

Remember that our application is battery-driven, so power consumption is a factor. In the sequential application, the processor is continuously working, even if the container is stationary. In our event-driven version, the processor is working only when something happens. In this instance, however, that will be every second, because this is the usual period that a GPS receiver will send a new National Marine Electronics Association (NMEA) position string. This means that for this basic version, there is not a lot to recommend one system over the other.

Now let's add some battery power savings. We will use the combination of the GPS information and the accelerometer to detect when the container is stationary and when it starts moving.

NOTE *Remember that acceleration is a change in velocity. You will get the same G-reading for the stationary container as you'll get when it's moving at a constant 30 mph (0G), because* *the accelerometer alone cannot detect when the container is stationary. Also, when a GPS receiver is stationary, the position data will continue to "wander" over an area of several meters, so it cannot be used alone.*

When the container is not moving, the GPS module will power down to an idle state. It will not send GPS data during this time and will draw considerably less power; the same goes for the GSM module. When movement is detected, both of these will be powered on again.

Suppose our container needs to be stationary for 10 minutes before the module goes into low power mode. We use a "moved" flag variable to record whether the container has moved. If the flag is false for 10 minutes, the low power mode will begin.

Our sequential code flow is as follows:

1. Poll the GPS receiver for the container's current position.

2. Calculate its position change.

3. If its position had changed, send data to the web service. Set the moved flag. Go back to step 1.

4. If the container's position has not changed, do the following:

    4.1 If moved flag is true, set to false; record current time; go back to step 1. If moved flag was false, go to step 4.2.

    4.2 Check current time: If stopped time is 10 minutes, go to 4.2.1; otherwise, go back to step 1.

        4.2.1 Devices are set to low power mode.

        4.2.2 In low power loop, check accelerometer.

        4.2.3 If no movement is detected, go back to step 4.2.1, else 4.2.4.

        4.2.4 If movement is detected, power up devices.

        4.2.5 Set the moved flag to true, and go to step 1.

Here's our event-driven code flow:

The event-driven model breaks the code into independent sections, allowing simpler and more thorough testing and simpler debugging. We add a timer, with an event every 30 seconds, a moved flag, and a **stopCount** variable.

1. When GPS data is received, any position change is calculated. If a change is noted, the application generates a **positionChanged** event.

2. On the **positionChanged** event, if the position change is greater than x meters, the application sends the new position to the web service. The moved flag is updated to true, and **stopCount** is reset to 0. If the position has not changed, the move flag is set to false.

3. **OnTimer Tick** is set to every 30 seconds; if the move flag is false, increment stop count. If **stopCount** is 20 (moved is false for 20 timer periods—10 minutes) generate **OnStoppedEvent**.

4. The **OnStoppedEvent** sets devices to low power mode.

5. If the accelerometer detects movement, devices are powered up if they're set to low power mode.

We have modified the application to detect whether the container has stopped moving for 10 minutes, put the GPS and GSM into a low power mode, and the application now monitors the accelerometer to detect movement. If the app detects that the container is moving, the devices are powered on and continue in normal mode until the container stops moving again. At this point, the stop detection is very simple, but if this were tested in the real world, the application would need to be far more comprehensive.

The point is that we have made a fairly simple change, but already the sequential code is starting to have loops within loops, while the event-driven mode has just added a few events. Each event and handler is an independent bit of code; any interaction from other events occurs via variables, so we maintain a separation between "rules" and "implementation." If we need to modify the rules, such as the amount of time the container is stationary or how we detect that the container is stationary, the event-driven model provides a simpler, more independent mechanism to do this, without a change in one area having an adverse effect in another area of code. This will be more and more difficult to achieve with the sequential model, as the complexity increases.

Now we have implemented responses to actions. The response is not dependent on how the action was generated or directly connected or coupled to the code that generated the event. We can now modify the rules that generate the event, independently of the code that handles the event.

Gadgeteer Application Flow

The .NET Micro Framework and Gadgeteer are event-driven frameworks and are modeled to make event-driven programming easy. However, there are many ways to implement a software solution to a problem, and some situations may well be better solved by a mixture of the two programming models. We will examine how we can implement both programming models into a Gadgeteer application.

The Gadgeteer Application Template

We can use the Gadgeteer-supplied Visual Studio Application Project template to create our new application. It will create the project and allow us to select the required mainboard and modules using the Visual Designer.

Our first step is to connect the module sockets to the correct mainboard sockets (as discussed in Chapter 2). The designer will generate code files as Program.generated.cs and Program.cs, and then add these to the project. It will also add all the required references to the Micro Framework, Gadgeteer, mainboard, and module libraries to the project. The Program.generated.cs file is controlled by the designer. It will hold the global instances of our mainboard and added modules. It will also initialize and connect the modules to the mainboard. This is implemented in the **Main()** method, which is the entry point to the application. The Program class is divided between the two files, Program.generated.cs and Program.cs. They are both parts of the same class but are spread across two files. Our code entry point is added to the Program.cs file.

NOTE *Do not modify Program.generated.cs; this file is maintained by the designer, and any changes you add here will get overwritten by the designer.*

The **Main()** method performs the following.

1. Creates the mainboard.
2. Creates the base application.
3. Initializes the designer-added modules.
4. Calls the **ProgramStarted()** method.
5. Calls **Program.Run()**, which starts the application.

The following code sample is from a typical Program.generated.cs file:

```
public partial class Program : Gadgeteer.Program
{
    // GTM.Module definitions
      Gadgeteer.Modules.Sytech.Serial2USB serial2USB;
      public static void Main()
      {
      //Important to initialize the Mainboard first
            Mainboard = new Sytech.Gadgeteer.Nano();
            Program program = new Program();
            program.InitializeModules();
            program.ProgramStarted();
            program.Run(); // Starts Dispatcher
      }
      private void InitializeModules()
      {
            // Initialize GTM.Modules and event handlers here.
            serial2USB = new GTM.Sytech.Serial2USB(2);
      }
    }
```

The following listing is from a designer-generated Program.cs file, before we added any of our own code:

```
public partial class Program
{
 // This method is run when the mainboard is powered up or reset.
 void ProgramStarted()
 {
 /************************************************************
 Modules added in the Program.gadgeteer designer view are used by
 typing their name followed by a period, e.g.  button.  or  camera.
 Many modules generate useful events. Type +=<tab><tab> to add a
 handler to an
    event, e.g.:
    button.ButtonPressed +=<tab><tab>
    If you want to do something periodically, use a GT.Timer and
    handle its
    Tick event, e.g.:
     GT.Timer timer = new GT.Timer(1000); // every second (1000ms)
     timer.Tick +=<tab><tab>
     timer.Start();
 *************************************************************/
```

```
// Use Debug.Print to show messages in Visual Studio's "Output" window
// during debugging.
    Debug.Print("Program Started");
  }
}
```

As mentioned, Gadgeteer/Micro Framework .NET applications are event-driven. Our designer-generated application will set up the environment, create and connect the modules, and hand them over to our application, so we can do any setup required (at step 4, the **ProgramStarted()** method). The next and final step is to start the execution loop (program.run), which is where our application code runs. This execution loop will run forever. Our application code needs to start operating after the main program execution loop is running.

The **ProgramStarted()** method, in the Program.cs file, is where we add any initialization we require, such as connecting up event handlers to the added modules. Remember that we are writing an event-driven application. (This method should have been called ApplicationInitialize, rather than **ProgramStarted**—I find the name misleading.) After application initialization, the **ProgramStarted()** method exits and **Main()** will now call **program.Run()**—starting the event-driven application framework. With the event-driven approach, we have connected up all our event handlers to the module events in **ProgramStarted()**. Our application flow is the code in the event handlers; it's how we respond to the actions triggers. As described in the event model, there is no actual start point for our application code, because it is non-sequential and it starts when the first event is fired.

Application Thread

The underlying application class program executes on the core Micro Framework thread. It is this thread that is used by the TinyCLR to manage the Framework. All the Window class and Windows Presentation Foundation (WPF) graphics components are controlled and rendered by this thread.

Gadgeteer components generate events, and it is a recommended practice for a module implementer to ensure that the event handler attached to the event executes on the main application thread. This ensures that if the handler code needs to update Windows graphic components, such as a text box, the update is called on the main thread. This is required for the graphic component to be reliably rendered to the screen. However, you need to keep in mind that the main thread is doing all the work in managing the underlying framework. If you start making large demands on its time by running long application methods on it, you will slow the underlying framework, or, in extreme cases, stop it from working.

If sections of code are running a tight loop—for instance, it might be constantly reading bytes from the serial port and then processing them to find the beginning and end of a message packet—these should run in their own thread. This will prevent them from putting an extra direct overhead on the main thread. If, as a result of doing some work, the code in this new thread needs to update a graphic component, the Gadgeteer library provides a simple mechanism to move the call over to the main thread. Running code in different threads is not a "magic" way to get the processor to do more work; ultimately, there is only one processor. All we are doing is splitting the execution tasks into sections and dividing the processor's time between these sections; it creates a fairer division of labor for the processor.

Creating a new thread is a fairly simple process. You create a new instance of the Thread class and initialize it with a delegate (a reference to a method). This method will run in the new thread. Once the thread is initialized, just call **thread.Start** to begin execution of the code in the thread. The following code sample shows how to create a new thread and then run some code in it in a loop. When the method used by the thread exits, the thread is terminated.

```
using Microsoft.SPOT;
using System.Threading;
namespace EmptyApp
{
    class threadDemo
    {
        private Thread m_thread;
        private bool m_threadRun;

        public threadDemo()
        {
            m_thread = new Thread(DoWork);
            m_threadRun = false;
        }

        /// <summary>
        /// Start the thread, and create a new
        /// one if required
        /// </summary>
        public void StartThread()
        {
            if (m_threadRun == false)
            {
                if ( m_thread == null)
                {
                    m_thread = new Thread(DoWork);
                }
                m_threadRun = true;
                m_thread.Start();
            }
        }

        /// <summary>
        /// Allows thread to be stopped
        /// </summary>
        public void StopThread()
        {
            m_threadRun = false;
        }

        /// <summary>
        /// This is our thread code section
        /// </summary>
        private void DoWork()
        {
            while ( m_threadRun)
            {
```

```
            // your code here
        }
        m_thread = null;
        Debug.Print("Thread has finished");
    }
}
```

In this example, we added a mechanism for allowing the thread to be stopped and also restarted. While the variable **m_threadRun** is true, the thread code will loop continuously. If we set **m_threadRun** to false, at the start of the next loop iteration, the thread will stop running.

The method **DoWork** will run on a different thread than the main application and will be "time-sliced" with all the other threads in the application. This gives you basic threading.

You should be aware of other aspects of threading, such as synchronization of variables between threads. For example, two threads might be trying to change a common variable, and one thread changes the value at a critical time for the other thread. To help avoid this, you can "lock" objects, so that only one thread can have access to them, causing another thread to wait until the lock has been released; this, however, is a more advanced topic than this basic introduction to threading. For more advanced usage of threads, refer to online C# tutorials or one of the many books on C# programming.

Classes and Project Code Files

I would recommend that you define areas of your application into classes, and always put a class into its own file. This will encourage a better coding style and lead to clearer, more manageable code. Do not try and put your entire application code into Program. cs. In fact, I would recommend that you put only your initial application set up in this file. Make your application a class, and use **ProgramStarted** to create the instance of your application class and pass in all the instances of the modules required.

Our example GPS tracker application has the following physical modules:

- Mainboard
- GPS module
- GSM module
- Accelerometer module

First, you'll need to define a class for each separate area of functionality: a class for GPS position handling, which uses the GPS and Accelerometer modules, and a class for communications, which uses the GSM module. Then define a class for your application that uses the position handling class and the communications class. This allows you to separate the different required components of your application and provides some abstraction from what the component does and how it does it.

The application needs to know whether the shipping container has moved and, if so, where to—it does not need to know, or be concerned about, how the position moved was determined. The position handling class only looks at how to determine whether the container has moved. If it has, it publishes an event—**OnPositionMoved**. It does not care who uses this information or what they do with it.

In **ProgramStarted**, create a new instance of your application class and pass it into the GSM module, GPS module, and Accelerometer module instances (created by the Gadgeteer designer).

In your application class, create new instances of the position handler, passing in the GPS and Accelerometer class instances. You also create a new instance of the communications class, passing in the instance of the GSM module.

Now hook up your event handlers in the application class for the position and communications events, and you are set to go. (Obviously, you also need to implement code for the handlers!)

Starting off your development with this kind of organized class structure will benefit you in the long term, helping you in developing the various aspects of the application and, more importantly, in debugging the application.

Figure 7-1 shows the class diagram for our project.

And here's the Program.cs code sample:

```
public partial class Program
    {
        private TrackerApp trackerApp;
        // This method is run when the mainboard is powered up or reset.
        void ProgramStarted()
        {
            trackerApp = new TrackerApp(gps, gsm, accell);
            trackerApp.Start();

            // Use Debug.Print to show messages in Visual Studio's
            // "Output" window during debugging.
            Debug.Print("Program Started");
        }
    }
```

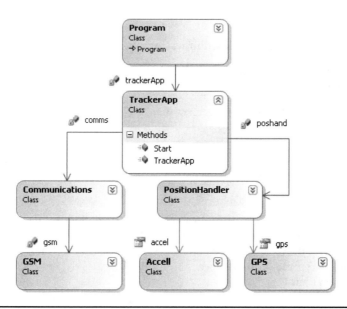

FIGURE 7-1 Tracker class diagram

Here's our TrackerApp.cs code:

```
public class TrackerApp
    {
        private PositionHandler poshand;
        private Communications comms;

        public TrackerApp(GPS gpsMod, GSM gsmMod, Accell accellMod)
        {
            poshand = new PositionHandler(gpsMod,accellMod);
            comms = new Communications(gsmMod);
            //now hook up event handlers here...
        }

        public void Start(){}
    }
```

Here's the **PositionHandler** code:

```
public class PositionHandler
    {
        public event EventHandler OnPositionChanged;
        public event EventHandler OnStationary;

        public PositionHandler(GPS gpsMod, Accell accellMod)
        {
            gps = gpsMod;
            accel = accellMod;
        }
        public void Start(){}
        public GPS gps{}
        public Accell accel{}
    }
```

And, finally, here's the **Communications** class code:

```
public class Communications
    {
        private GSM gsm;
        public Communications(GSM gsmMod)
        {
            gsm = gsmMod;
        }

        public void SendPosition(){}
    }
```

Now we can work our way through each class and add the implementation code. Each class can be worked on and tested independently. You can make a dummy test class that fires the events for the **PositionHandler** to test the TrackerApp application. In the **PositionHandler** you can make a dummy GPS class to generate test data to test just the position handler.

Using Sequential Application Code

As promised, here is how you can combine sequential code and event-driven code in a Gadgeteer application. If the main loop of our application is modeled around a sequential application model, we need to find a suitable entry point for our program loop. The main loop may be sequential, but there could be some modules connected that modify how the sequential loop operates—for instance, when a button is pressed. So we can use a mixture of sequential and event-driven models.

Here is the catch: **ProgramStarted()** needs to execute and exit completely, so the **Main()** method can then call **program.Run()**; otherwise, the application framework is not running. For a completely event-driven application (the norm), this is simple. If our application has a sequential general flow, our application is in a loop that then does a number of things, and usually forever. If we put that code here, we would never exit **ProgramStarted**. So we cannot put the entry to our application here, because it could prevent **program.Run** from being called. We need to start after **program.Run** is called.

If your application is totally driven from events attached to the modules, this is not a problem. But if there is a sequential flow to the beginning of your application before it becomes event-driven, you need to delay the start of this sequential code until after the application has started.

Here's an example of an application with a sequential flow:

1. Do function A.

2. Do function B.

3. Do function C.

4. Repeat from Step 1.

If we place this code in **ProgramStarted**, it would look like this (which will not work correctly!):

```
public partial class Program
    {
        // This method is run when the mainboard is powered up or reset.
        void ProgramStarted()
        {
            button.ButtonPressed += new
    Button.ButtonEventHandler(button_ButtonPressed);
            button.LEDMode = Button.LEDModes.ToggleWhenPressed;
            while (true)
            {
                DoFunction1();
                DoFunction2();
                DoFunction3();
            }
            // we will never get past this point

            // Use Debug.Print to show messages in Visual Studio's
            // "Output" window during debugging.
            Debug.Print("Program Started");
        }
        void button_ButtonPressed(Button sender, Button.ButtonState
            state)
```

```
    {
        Debug.Print("Button Pressed");
    }
    void DoFunction1()
    {}
    void DoFunction2()
    {}
    void DoFunction3()
    {}
}
```

As you can see, this code would never exit **ProgramStarted**, so **Main()** would never call **program.Run()**. This means the Micro Framework application layer would not get started. The underlying operating system does not get started, so the button events are not active. The button or any other Micro Framework event would never get fired. In effect, you are not running Gadgeteer (or full Micro Framework).

In this situation, you need to move the sequential loop so it is executed after the application main thread has started. One way to do this is with a timer. The timer is set for a one-time delay, starting from **ProgramStarted**. The Tick handler contains the loop. This allows **ProgramStarted** to exit and **Main()** to call **program.Run**. After the delay, the timer Tick event is fired (by now **program.Run** has been called) and calls our sequential code. However, there are still some subtle points that could spell failure here—the main one being that you need to be running your sequential loop code on a different thread to the Gadgeteer application. The Gadgeteer timer "tick" runs in the main thread.

First, you must not take over the main application thread. If you put the main thread into a tight loop, you can lock out the Micro Framework operating system. It will not be able to get enough use of the thread, as your code has it locked in a tight loop. Putting the odd **Thread.Sleep(x)** in your code will not help, because when this thread is asleep, no one gets to use it (you are putting the main thread to sleep). You need to run your sequential code on a different thread from the main application. (This is covered in the "Application Thread" section of this chapter.)

There are actually two conditions required for running your sequential code:

- Execution of sequential code must start after **program.Run** is called.
- Your sequential code must run on a thread different from that of the main application.

If you use a timer to start your code after **program.Run**, *do not use* a Gadgeteer timer; instead, use a **System.Threading** timer. The Gadgeteer timer will marshal (move) the Tick event handler onto the main application thread, but you need to run the code on a different thread. If you use a .NET timer, the Tick handler will be run on a new thread that's different from the main application thread.

However, a simpler answer is just to create a new thread in your sequential application and run your code in a loop from this thread. Then it does not matter where you start it from, even in **ProgramStarted**, because the sequential code is running in its own thread, and it will not block the main thread, allowing **ProgramStarted** to exit.

We have moved our sequential application code into its own class: **SequentialApp**. Putting your code into classes and the classes in their own file is good practice; as your code gets more complicated, this will make your debugging simpler. It also promotes reuse of code across different projects by allowing you to share code. Always avoid putting everything in one big file.

When we call **StartApp**, we now create a new thread and execute our tight loop in that thread.

The following code listing is an example of implementing a sequential application in its own thread:

```
public partial class Program
    {
        private SequentialApp m_app;

        // This method is run when the mainboard is powered up or reset.
        void ProgramStarted()
        {
            button.ButtonPressed +=
            new Button.ButtonEventHandler(button_ButtonPressed);
            button.LEDMode = Button.LEDModes.ToggleWhenPressed;
            m_app = new SequentialApp();
            m_app.StartApp();

            // Use Debug.Print to show messages in Visual Studio's
            // "Output" window during debugging.
            Debug.Print("Program Started");
        }

        void button_ButtonPressed(Button sender, Button.ButtonState
          state)
        {
            m_app.ButtonPressed();
        }
    }
```

This code listing is the dummy sequential application code class:

```
public class SequentialApp
    {
        private bool test;
        private Thread m_thread;
        public void StartApp()
        {
            m_thread = new Thread(dowork);
            m_thread.Start();
        }

        private void dowork()
        {
            while (true)
            {
                DoFunction1();
                DoFunction2();
                DoFunction3();
            }
        }
```

```
public void ButtonPressed()
{
    // change application flow if button pressed
    Debug.Print("Button Pressed");
}
void DoFunction1()
{}
void DoFunction2()
{}
void DoFunction3()
{}
}
```

The **SequentialApp** class is created in **ProgramStarted**, and then **m_app.StartApp** is called. This will create a new thread—**DoWork()**; then the infinite loop is executed on this thread. Because this is a different thread, it does not block the **StartApp** call. After **m_thread.Start()** is called, **StartApp** returns, allowing **ProgramStarted** to exit and **Main()** to start the main application thread. Meanwhile, our sequential loop is executing in its own thread. Now when the button is pressed, the **SequentialApp** is informed by the call to **SequentialApp** in the button press handler.

Note that this handler, even though it is in the **SequentialApp** class, will run on the main thread, as it was triggered by the Gadgeteer main thread.

Summary

In this chapter, we have explored how a Gadgeteer application is constructed and how we should structure and design our application code. You have seen how Gadgeteer applications are event-based, not sequential. We have looked at simple threading and the threading structure of Gadgeteer. The rest of Part II offers practical examples of simple Gadgeteer projects and explores the various aspects of using modules.

CHAPTER 8

Data Input and Output Projects

I n this chapter will we look at using data input and output modules in several projects. In general, these projects will use general purpose input/output (GPIO), analog, Inter-Integrated Circuit (I2C), and Serial Peripheral Interface (SPI) interfaces to the mainboard.

From an application point of view, the interface used by the module isn't very relevant. For instance, an accelerometer module could be designed to use an SPI interface or an I2C interface, and the application will use the high-level interface, which simply exposes the G (acceleration) data for each axis. The only thing relevant to the interface between modules and mainboards is whether the mainboard has the required physical resources to allow the modules to be used. All of the low-level interface is implemented in the driver, so the application sees only the top-level interface.

We will explore designing a complete application in a structured manner, allowing a more flexible approach to testing and extending the complete application as the requirements change.

We will look at using a selection of these modules individually. Each module or sensor will have its own project, to develop a "reusable" component class. Then we will generate our final project using our reusable components classes together in one system. We'll build the code for each module first (in its own project), and then show how code can be reused to generate the final combined project. All the separate projects will be combined into one Visual Studio solution. For each project, we will explore an aspect of programming using the Micro Framework, .NET, and the Gadgeteer framework.

We start by generating an empty solution in Visual Studio to hold our projects.

Create an Empty Solution in Visual Studio Express

Visual Studio Express differs from the "premium" versions in the way it saves new projects. Saving occurs at the end of the creation process rather than the beginning. It does not supply a template for creating an empty solution (a *solution* is a container for a number of projects) in a default installation. When a new project is created, Express creates a default solution with that project's name. We want to create an empty solution with a solution name that differs from that of any of our projects. This helps enforce our "structured" method of writing code for large projects. In the "premium" editions of Visual Studio this is simple, there is a template for an empty solution, but in Visual Studio Express we need to jump through a few hoops to achieve this. There is a template for an Empty Project, which will create a Solution and put a dummy project in the solution. We will use this template to create a Solution with the name we require and delete the default project, with the same name, and add our own projects. This will give us the structure we want, a Solution (container for our projects), and the group of projects for this task. This way we keep our development work organized. The Solution name is the descriptive "title" of our work and contains all the elements.

Let's start by creating a new solution to provide a container for our new projects.

1. Open Visual Studio and select New Project:

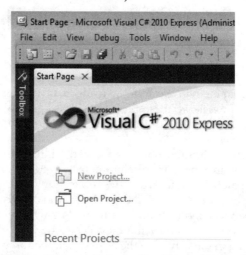

2. In the Visual Studio New Projects dialog, select Empty Project from the Installed Templates section, on the left. Type **Chapter8** in the Name field, and then click the OK button as shown next:

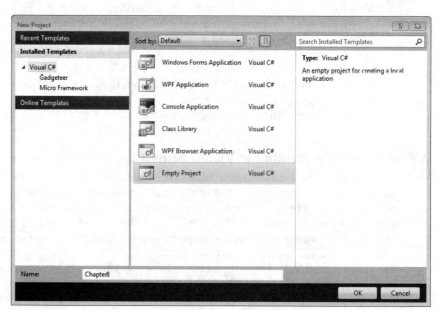

The new solution, Chapter8, is created and can be seen in the Solution Explorer, as shown next. The template has created a default project with the Solution name 'Chapter8'. What we really want is an empty Solution, i.e., no projects, so we will delete this default project.

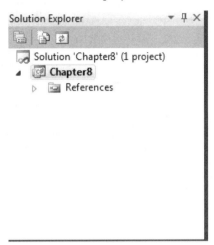

3. Visual Studio Express will not let us delete the only project in this solution. So we must first add our first *real* project, which will be the SPI Display Demo. In Solution Explorer, right-click Solution "Chapter8" (1 project) and select Add | New Project, as shown next:

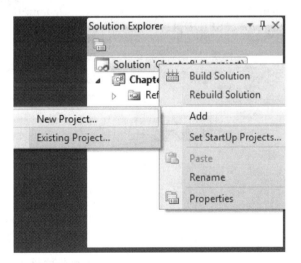

4. Visual Studio will display a warning dialog (shown next) saying the current project must be saved first. Click Save.

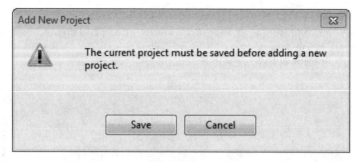

5. In the Save Project dialog, click the Browse button to navigate to the folder where you want to save the project. Make sure the "Create directory for solution" checkbox is checked—this creates a solution folder and then puts the projects in their own subfolders within the solution folder. Then click Save.

6. In the Add New Project dialog, you can see the project location at the bottom of the dialog. Select the Gadgeteer template in the Installed Templates area, and then click the .NET Gadgeteer Application template in the middle pane. Name the project **SPIDisplayDemo**, as shown next:

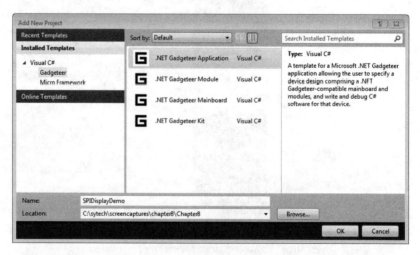

7. The new project will be added to the solution. Now we can delete the original "dummy" project. In Solution Explorer, right-click the Chapter8 project (not the solution) and select Remove.

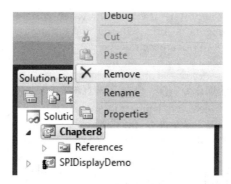

8. A new warning dialog will warn you that the project will be removed. Click OK.

Now, in Solution Explorer, our new solution, Chapter8, and our new Gadgeteer project, SPIDisplayDemo, appear:

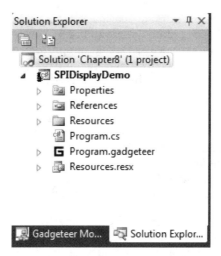

Creating a solution with a collection of related projects promotes writing code in an orderly fashion and saves the code in separate project folders, grouped under a solution folder. Giving the solution and the projects descriptive names helps document them and reminds you of what they contain when you return to them at a later time. It is also the first step in generating a reusable code base.

Now let's add our new projects.

SPI Display Module Using Project Resources

For our SPI project, we'll use the Seeed OLED (Organic Light Emitting Device) SPI Display module, but you could use any SPI display module. This module has a small display with a resolution of 128 × 128 pixels (picture elements).

The module is interfaced to the mainboard using SPI. The module driver handles all the low-level details of controlling the display. The firmware will also expose a Gadgeteer Simple Graphics interface, which will convert the raw picture data into a format compatible with the display, intercept the base Micro Framework "draw" routines, and redirect the graphics image data to the display. This allows mainboards that do not natively support LCD displays to use a display.

NOTE *Ideally, the mainboard would support the bitmap conversion routines in native C/C++ code. This is an extension required by Gadgeteer. If the board does not support bitmap conversion in native code, a managed C# conversion function is supplied by Gadgeteer. However, this managed C# code will be very slow, so it's suitable only for simple text displays.*

In this project, the module screen will display a line of text, and then, 1 second later, the screen will update with an image. The text string, font, and the image file displayed will be stored with the other project resources in the Resources directory and loaded from that directory.

When we add an item to the Resources directory, such as an image, the actual binary data of the file is stored in the main project dynamic link library (DLL), rather than being stored as an external file that has to be deployed to the device. So the file is actually "embedded" into the program file. This allows a more integrated approach to storing display data, such as text and graphic images.

We will use our display in our final project to display data. We have already added the project code outline in the earlier procedure, so we can now start implementing the code for the project.

NOTE *In the following examples we will use modules from GHI and Seeed. You will need to have installed the module SDKs for each of these manufacturers to add the modules to the designer toolbox. They can be downloaded from the Gadgeteer support page of each manufacturer's website. The URl for the GHI website is www.ghielectronics.com and for Sytech Designs it is www.gagdeteerguy.com.*

Open the designer canvas by double-clicking on the "Program.gadgeteer"' file, in the Solution Explorer SPIDisplayDemo project. In the designer canvas, add a **Seeed .OledDisplay** module and a mainboard—in this example, we've used a Sytech NANO mainboard. Connect the display to a valid socket on the mainboard, as shown in Figure 8-1.

Seeed.OledDisplay

oledDisplay

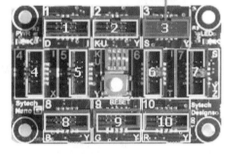

FIGURE 8-1   Designer canvas

NOTE *The instance name of any modules you add to the designer are displayed beneath the module. In our example, the OLED display is called* `oledDisplay`. *This is the name we use in our code to refer to the instance of the module—for instance,* `oledDisplay .SimpleGraphics`.

Adding Project Resources

Now let's add a JPEG image to display on the module. Any image will do; we can resize it to fit the size of the module display, which, in our example, is 128 × 128 pixels. For this example, we'll use a JPEG image of a clownfish. To add the JPEG to the Resources directory, follow these steps:

1. In Solution Explorer, right-click the Resources.resx file and choose Open from the context menu.

2. In the Visual Studio Resource Editor menu bar, select Add Resource | Add Existing File, as shown next.

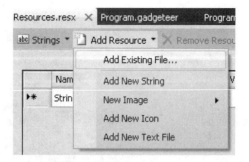

3. In the File dialog, browse to the JPEG file you want to add and click Open.

The Resource editor will copy your file to the project's Resources directory and add it to the resources. The image will then appear in the resource editor window:

To add a string to the resources, follow these steps:

1. In Solution Explorer, double-click or right-click Resources.resx and choose Open to open the resource editor.

2. Select Add Resource | Add New String from the menu bar to open the string resource page, where all the strings are presented in a grid. The first column is the name you use to refer to the string, the next column is the actual string, and the rightmost column allows you to enter a comment to describe the string function. We will add a "Hello World" string, as shown next:

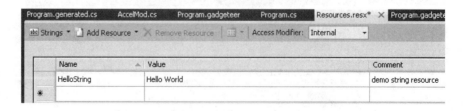

3. Now to add the code for our example. This is just a simple demo, so we will put all the implementation in the Program.cs file; we will not be reusing this code in later projects.

Our code will do the following on the display module:

1. Clear the display.
2. Display a "Hello World" string.
3. Display the fish image one second later.

We will use a Gadgeteer timer to create the one-second delay and use a tick handler to display the image to the screen. Here's the complete code listing of the modified Program.cs file:

```
using Microsoft.SPOT;
using GT = Gadgeteer;
using Timer = Gadgeteer.Timer;

namespace SPIDisplayDemo
{
    public partial class Program
    {
        // This method is run when the mainboard is powered up or reset.
        void ProgramStarted()
        {
            // step 1 - clear display
            oledDisplay.SimpleGraphics.Clear();

            // step 2 Display a simple line of text
            string text =Resources.GetString(
                Resources.StringResources.HelloString);
            oledDisplay.SimpleGraphics.DisplayText(text,
                Resources.GetFont(Resources.FontResources.small),
                GT.Color.Red,10,10);
            // step 3 set up our 1 second timer
            GT.Timer timer = new Timer(1000,
                                        Timer.BehaviorType.RunOnce);
            timer.Tick += new Timer.TickEventHandler(timer_Tick);
            timer.Start();

            // Use Debug.Print to show messages in Visual Studio's
            //"Output" window during debugging.
            Debug.Print("Program Started");
        }

        /// <summary>
        /// Our timer tick handler
        /// </summary>
        /// <param name="timer"></param>
        void timer_Tick(Timer timer)
        {
```

```
// Display our fish image from the resources
oledDisplay.SimpleGraphics.DisplayImage(
    Resources.GetBitmap(Resources.BitmapResources
    .fishJPG),0,0);
    }
  }
}
```

We used the Gadgeteer **SimpleGraphics** interface to access the display, to write a line of text, and to display the image. The Resources directory contains information to define the font to use in writing the text string, the text string itself, and the image file.

Using **SimpleGraphics**, we can write text to the screen and pass in the text string, font, color, and pixel position for the location of the first character of the text. It also allows an image to appear on the display. We also indicated the pixel coordinates of the top-left corner of the image to indicate where it should be drawn on the display. The font was automatically added by the project template. We used the .NET **Resources** class to access our resources.

Now build the project, deploy it, and debug. The application will run on the hardware, displaying the text string "Hello World" at the tenth pixel row and tenth pixel column of the display. After one second, the fish image will be sent to the display.

All this happened over an SPI interface from the mainboard to the module. The application running on the module is shown in Figure 8-2.

FIGURE 8-2 Application running on the module

I2C Accelerometer and Process Data Threads

We'll now add a new project, an I2C Accelerometer application, to our solution. We'll implement the accelerometer code as a reusable class so that we can use the class in other projects. This project will be a "test bed" for our new accelerometer class. In the final project of the chapter, you'll learn how to reuse our code in a larger project.

In this project, we will also explore a *data input type class*. Basically, this class gets its data from some external source, such as a sensor. It's then required to process that data in some manner—perhaps via some digital filtering or looking for specific values and reacting to them. It then makes the processed data available for use by other application classes. In short, the class gets the data, does some work, and presents the result.

This process will most likely be achieved with a *repetitive-loop*. We'll look at how to use a separate thread to do the data processing and how to make the resulting data "thread safe." We'll get the raw data for each axis of the accelerometer, and then average the samples over a period of time. Our processing thread will be accessing and updating the samples into an array. Our data output function will also access the data sample array. But the output function is called by our application, and the caller will be on a different thread to our internal processing. We do not want the calling thread to access the data array while our internal thread is updating it. We want to "lock" the array processing, so only one thread at a time has access to it. This is a very common scenario in embedded device design.

Let's get started on our new project.

1. In Solution Explorer, right-click the Solution "Chapter8" project and choose Add | New Project.

2. In the Add New Project dialog, in the Installed Templates pane, select the Gadgeteer templates, select .NET Gadgeteer Application in the middle pane, and name the project **AccelDemo**. Then click OK.

3. In the designer canvas, add a mainboard and a Sytech Accel3Axis module. I am using a NANO mainboard in this example. Connect up the Accel3Axis module to a valid mainboard socket. The designer canvas will look similar to Figure 8-3.

4. In Solution Explorer, right-click the AccelDemo project and choose Add | Class.

5. In the Add New Item dialog, type **AccelMod.cs** for the name of the new class. Click Add to create the new class and add it to the project.

AccelMod.cs is our reusable **Accelerometer** class. In this class, we will initialize the module's I2C connection, set its mode to measurement, and set the G range to 2G. We will then create a public method to return the current X, Y, and Z data. Our example is very simple and will do some simple data processing on the raw data. We'll get the data at a regular rate and then average the samples over a ten-cycle period.

The code is designed so that the number of samples to average can be easily changed. Currently, we are using ten samples, but you can change the value of the constant **SAMPLES** to change the number of samples used. You could implement other data processing functions, such as calculating the current angle in the X and Y axes, and whether the board is inverted or right side up (Z axis). Our aim is to demonstrate how we can implement some behavior in the module and make the results available for other classes to use.

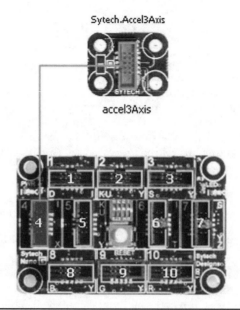

Sytech.Accel3Axis

accel3Axis

FIGURE 8-3   AccelDemo project Designer

We'll demonstrate using a separate thread to perform our processing and making the data result thread safe using a locking object. All our processing will be on a thread different from the main application thread. (Another approach would have been to use a system timer, but we want to demonstrate the thread approach here.)

Here's the complete code for the class:

```
using System.Threading;
using Gadgeteer.Modules.Sytech;

namespace AccelDemo
{
    public class AccelMod
    {
        private Accel3Axis m_accel;

        //data averaging arrays
        short[] m_xAxis;
        short[] m_yAxis;
        short[] m_zAxis;

        private uint m_avgOffset;
        //change this value to set number of samples
        private const uint  SAMPLES = 10;

        //thread variables
        private object m_lock;
        private Thread m_processThread;
```

```
/// <summary>
/// Constructor
/// Pass in the accel module
/// </summary>
/// <param name="?"></param>
public AccelMod(Accel3Axis accel)
{
    m_accel = accel;
    InitModule();
}

/// <summary>
/// Initialize the module
/// </summary>
private void InitModule()
{
    //create our data arrays for averaging
    m_xAxis = new short[SAMPLES];
    m_yAxis = new short[SAMPLES];
    m_zAxis = new short[SAMPLES];
    m_lock = new object();
    m_avgOffset = 0;
    //set up physical module
    m_accel.InitI2C();
    m_accel.SetMode(Accel3Axis.Mma7455Mode.modeMeasurement,
        Accel3Axis.Mma7455gSelect.g2);
    // create and start our work thread
    m_processThread = new Thread(Process);
    m_processThread.Start();
}

/// <summary>
/// Get the average over x samples
/// of the x axis data.
/// </summary>
/// <returns></returns>
public short GetXdata()
{
    return CalcAvgData(m_xAxis);
}

/// <summary>
/// Get the average over x samples
/// of the y axis data.
/// </summary>
/// <returns></returns>
public short GetYdata()
{
    return CalcAvgData(m_yAxis);
}

/// <summary>
/// Do averaging of an axis data array
/// Thread safe use of data arrays
/// </summary>
```

```
/// <param name="dataArray"></param>
/// <returns></returns>
private short CalcAvgData(short[] dataArray)
{
    short dataCalc = 0;
    lock(m_lock) // thread safe critical section
    {
        for (int offset = 0;offset < SAMPLES;offset++)
        {
            dataCalc += dataArray[offset];
        }
    } // end critical section
    dataCalc /= (short)SAMPLES;
    return dataCalc;
}


/// <summary>
/// Process data thread loop
/// </summary>
private void Process()
{
    short x = 0;
    short y = 0;
    short z = 0;

    while(true) // run forever
    {
        if (m_accel.ReadAll(ref x, ref y, ref z))
        {
            lock (m_lock)
            {   // critical section - locked
                m_xAxis[m_avgOffset] = x;
                m_yAxis[m_avgOffset] = y;
                m_zAxis[m_avgOffset] = z;
            } // lock is freed here
            m_avgOffset++;
            if (m_avgOffset >= SAMPLES)
            {
                m_avgOffset = 0;
            }
            // now put the tread to sleep for 60mS
            Thread.Sleep(60);
        }
    }
}
```

We pass the sensor into the constructor of the class and then initialize everything. It's always a good idea to put your initialize tasks into a separate function. We do any setup required by our sensor and the processing functions. In this example, we create our processing thread and set up and create the arrays required for our averaging function. We have coded the setup to use a constant, **SAMPLES**, that defines the number of samples to average. This adds flexibility to the code by allowing the number of samples to be changed by setting the value of the constant.

Our processing thread gets the raw data from the sensor and places the "sample" into the next available space in the data array. When we get to the end of the array, we loop back to the first position. This way, we always have the last "x" samples in the array. The data arrays are also used by the function that calculates the average data and returns it to an external calling function. Because this caller will be on a different thread, we control access to the arrays, so that only one thread has access at a time. We do this with a .NET lock function. We define an object to act as the token. The "critical" section of code controlled by the lock is located between the {} braces:

```
lock(m_lock)
{
       code ....
}
```

When a thread requires access to these "critical" sections, it requests the token (m_lock). If the token is not already in use, it is handed over to the requesting thread, and it then has ownership. If the token is in use by another thread, the calling thread is made to wait until the token is free.

NOTE *Be careful to ensure that you do not end up in a deadlock situation. In a deadlock, thread A has the token but is waiting for thread B to release another token. Meanwhile, thread B has the token required by thread A but is waiting for the token that thread A is holding—they will both wait forever!*

At the end of our processing thread loop, we put the processing thread to sleep for 60mS, so effectively our data is being pulled from the sensor every 60mS. While the thread is asleep, it does not use any of the processor time.

Our data is made available by the **Get[X/Y]Data** functions, which use a common function to sum the array data and divide it by the number of samples. While the data arrays are accessed, the token is required. Putting this data handling and use of the lock token in one common function helps you keep track of the lock and manage any possibility of thread lock. Having the token used throughout the class can make it difficult to keep track of the token, from a programming point of view.

Our application is very simple and is used to test our new sensor class. We create an instance of our new sensor class (**AccelMod**) and pass in the hardware module. We use a timer to generate a tick event every second. In the tick handler, we get the X and Y axes averaged data and display the values on the debug screen. Here's the code for the Program.cs file:

```
using Microsoft.SPOT;
using GT = Gadgeteer;
using Timer = Gadgeteer.Timer;

namespace AccelDemo
{
    public partial class Program
    {
        private AccelMod m_accell;
```

```
// This method is run when the mainboard is powered up or reset.
void ProgramStarted()
{
    // create our sensor class
    m_accell = new AccelMod(accel3Axis);

    GT.Timer timer = new Timer(1000);
    timer.Tick += new Timer.TickEventHandler(timer_Tick);
    timer.Start();
    // Use Debug.Print to show messages in Visual Studio's
    //"Output" window during debugging.
    Debug.Print("Accel Test Started");
}

void timer_Tick(Timer timer)
{
    short xdata = 0;
    short ydata = 0;
    xdata = m_accell.GetXdata();
    Debug.Print("X Data :" + xdata);
    ydata = m_accell.GetYdata();
    Debug.Print("Y Data :" + ydata);
}
    }
}
```

Next is a sample of the data sent to the Visual Studio Output window, with the application running:

```
Using mainboard Sytech Designs Ltd Nano version 1.0
Accel Test Started
X Data :-16
Y Data :-31
X Data :-17
Y Data :-31
```

In this project, we covered the basics of writing an application-level sensor class that uses its own processing thread. We have also looked at using *locks*, to synchronize access to the data from different threads. The class has been written so that it can be reused in other projects.

Gadgeteer DaisyLink

Eventually, as you work, you'll find you don't have enough sockets on your mainboard. If your mainboard has only three I2C sockets and you want to connect four I2C modules, you have a problem. The answer to this problem is DaisyLink, a protocol defined by Microsoft Research for use with Gadgeteer. It allows modules to be daisy-chained—that is, you can connect module 1 to the mainboard and connect module 2 to module 1, module 3 to module 2, and so on.

DaisyLink basically extends the I2C protocol and automatically works out which messages are for which module. DaisyLink modules are "active," so they require their own onboard processor to implement the DaisyLink protocol. This involves more work than simply designing a module using ICs (integrated circuits) that already incorporate standard protocols, such as I2C and SPI. Currently, few DaisyLink modules are available, but this could change in the future. GHI Electronics manufactures a DaisyLink development board that lets you design and prototype your own DaisyLink modules, but it requires that you are able to write code for a Cortex M0 microprocessor.

The DaisyLink protocol lets you use any mainboard general purpose input/output (GPIO) type socket, as long as it supports type X, GPIO 3. Ideally, the DaisyLink protocol needs to be implemented on your mainboard in native code (C\C++), but not all mainboards support this. If the mainboard does not support a native protocol implementation, Gadgeteer provides a managed C# implementation of the protocol. As long as your mainboard is reasonably fast, this managed implementation works fine.

For our next project, we'll use DaisyLink to link two multicolor LED DaisyLink modules. Let's get started.

1. Add a new Gadgeteer project to your Chapter8 solution, as described earlier in the chapter. Name the project **DaisyLink**.

2. In the designer canvas, add a mainboard and two DaisyLink multicolor LED modules.

3. Connect the first module to any valid mainboard socket, and then connect the second LED module to the first, as shown in Figure 8-4.

   Note how the designer has named the first module you placed *led* and the second one *led1*. You'll use these instance names to select the module required in your code. We will continue our example of implementing the code in a separate class, even though it is very simple. We'll use this class in our final project.

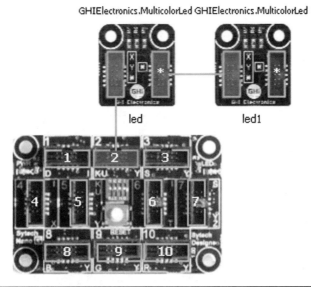

Figure 8.4   DaisyLink designer canvas

On startup, turn on each LED. The first module will be green and the second red. We've have added two methods to allow you to set the flashing LED in each module and also set the color.

4. In Solution Explorer, right-click the project and choose Add Class. Name it **LedDisplay**. The code for the class is shown next; it is pretty simple and self-explanatory:

```
using Gadgeteer.Modules.GHIElectronics;
using GT = Gadgeteer;

namespace DaisyLink
{
    public class LedDisplay
    {
        private MulticolorLed m_led1;
        private MulticolorLed m_led2;

        public LedDisplay(MulticolorLed led1, MulticolorLed led2)
        {
            m_led1 = led1;
            m_led2 = led2;
            Initialize();
        }

        private void Initialize()
        {
            //control the first LED module
            m_led1.TurnColor(GT.Color.Green);
            // control the 'linked' module
            m_led2.TurnColor(GT.Color.Red);
        }

        public void BlinkLed1Color(GT.Color color)
        {
            m_led1.BlinkRepeatedly(color);
        }

        public void BlinkLed2Color(GT.Color color)
        {
            m_led2.BlinkRepeatedly(color);
        }
    }
}
```

Our program class will create an instance of our new class, passing in the LED modules, and it will then set up a timer to run once after 1 second. The handler for the timer tick will set each of the LED modules to flash the LEDs a different color. The code is shown here:

```
using Gadgeteer;
using Microsoft.SPOT;
using GT = Gadgeteer;
```

```
namespace DaisyLink
{
    public partial class Program
    {
        private LedDisplay m_display;

        // This method is run when the mainboard is powered up or reset.
        void ProgramStarted()
        {
            m_display = new LedDisplay(led,led1);

            // timer will fire once after 1 second
            GT.Timer timer = new Timer(1000,Timer.BehaviorType
            .RunOnce);
            timer.Tick += new Timer.TickEventHandler(timer_Tick);
            timer.Start();

            // Use Debug.Print to show messages in Visual Studio's
            //"Output" window during debugging.
            Debug.Print("Led test Started");
        }

        private void timer_Tick(Timer timer)
        {
            Debug.Print("change LEDs to flashing");
            m_display.BlinkLed1Color(GT.Color.Red);
            m_display.BlinkLed2Color(GT.Color.Blue);
        }
    }
}
```

If you build and run this, you will see both LEDs light up—one green and one blue. Then after 1 second, the timer fires, changes the color of both LEDs, and starts them flashing.

TIP *If you find that blue and green are reversed on your LED modules, a call to the property* [MulticoloredLed].GreenBlueSwapped = ![MulticoloredLed].GreenBlue Swapped *will fix this (using the instance name of your module).*

A Combined Module Project

In the final project of the chapter, you'll learn how to reuse code from earlier projects in a larger, more complex project. This allows you to break down your project into individually testable sections. At this point, you know all the elements of your project work, so you can concentrate on getting them to work together to achieve your final design requirements.

This project will demonstrate one of the principal aims of Gadgeteer. All of the projects so far in this chapter have used a Sytech NANO mainboard. We will demonstrate the "modular" capability of Gadgeteer by using a different mainboard in our final project. We'll use the code from the AccelDemo and DaisyLink projects and integrate these modules into our new project, which will use a GHI Hydra mainboard. We'll use modules from three different manufacturers, and no changes are required to the original code classes to do this. As long as the mainboard supports the Gadgeteer interface sockets required by your module code, it will work with the mainboard.

You'll learn how to separate your application from the Gadgeteer-generated application and add the different elements required for your application to their own classes. This is essential for more complex projects, because it helps you maintain a testable and bug-free development process.

TIP *Putting all the code into the Program.cs file is not the way to go.*

Our project will use a combination of input devices that generate events that will trigger a response on an output device. We'll make a new input device sensor using a joystick that will act as a two-channel switch. We'll look at the position of the joystick X and Y axes: the X axis will be one channel and the Y axis will be the other. If an axis's joystick position is greater than halfway (center position is halfway), then we will interpret that channel as on, and below center will be off. The joystick position will be actioned when we press the joystick button; we'll look at the position of the joystick and determine the on/off state of each of the two channels. We'll then fire a custom event, **OnInputChange**. This can notify any "interested" class that the input data is available.

The class diagram of our project is shown in Figure 8-5, which shows the structure of the project and the relationship between the classes.

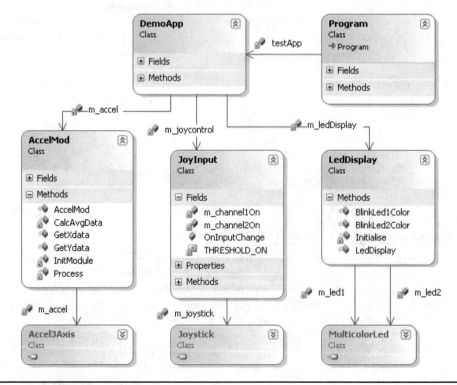

FIGURE 8-5 Project class structure

We'll use the **JoyInput** class to control two multicolor LED modules. Each data channel for the **JoyInput** will control the color of the flashing LED on the multicolor LED board. We will use the code we developed in the DaisyLink project, as is, for the LED displays. These functions are event-driven. When the joystick button is pressed, we generate an event, and this triggers reading the channel values and sets LED colors.

At the same time, we'll use an accelerometer module as another sensor input; every second we will display the G values for X and Y to a SPI OLED display. We will use a timer to trigger the events once every second.

Although the code does not do anything particularly useful, it demonstrates handling multiple independent sensor inputs and their data changes, while maintaining an orderly code structure that can be easily extended. It is also an example of an event-driven application.

Create the Project

Add a new Gadgeteer project to your Chapter8 solution. Name the project TestApplication. In the designer canvas, add the following from the Toolbox:

- FEZ Hydra mainboard
- GHI joystick
- Seeed OLED display
- Sytech Accel3Axis
- Two GHI multicolor LED modules

Figure 8-6 shows the modules and the socket connections we have used. The modules can be connected to any valid socket; you don't have to copy these connections exactly.

Now we'll implement the project.

1. Right-click the project name and choose Add | New Folder. A new folder will be added to the project, named New Folder and highlighted in the dialog. Type in the name **Application** in the highlighted area. We will put all our code classes in this folder.

2. Right-click the new Application folder and select Add | Class. Name the class **DemoApp**. This will be our application class. We will reuse the modules we developed earlier, **AccelMod** (from the AccelDemo project) and **LedDisplay** (from the DaisyLink project).

3. Open the AccelDemo project and right-click and hold on the AccelMod.cs file; you can now drag-and-drop this file into the Application folder of our new project. This will make a copy of the file and place it in our new project. Do the same for the LedDisplay.cs file from the DaisyLink project.

4. We need one more new class. We are going to use a Joystick module as a two-position switch. We need a class to define this behavior. Right-click the project and select Add | Class. Name the class **JoyInput**.

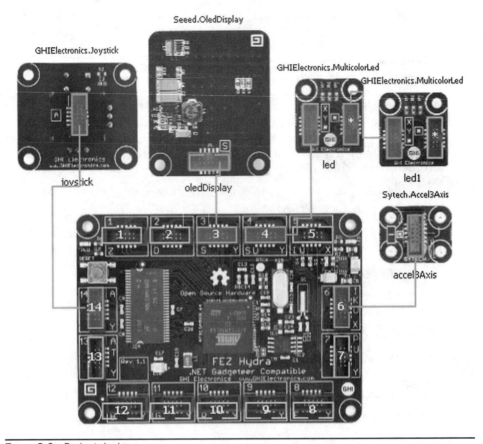

FIGURE 8-6 Project designer

In Solution Explorer your new project should appear as shown here:

JoyInput Class with Event

This part of the project will demonstrate using a sensor input module, processing its input data, responding to one of the module's events, and creating our own custom event for our sensor class.

We will use the two analog inputs of the joystick to create a two-channel switch. When an analog input is above the center position of the joystick axis, we will interpret that as on; if it is below or centered, we will treat that as off. The joystick module passes back a joystick axis (X or Y) position as a decimal number between 0 and 1, where 0.5 is (approximately) the center position. We will compare the input value with a constant 0.6; if the value is greater than 0.6, we treat the input as on, and if below 0.6, it will be considered in the off position.

The joystick class has a "button" event, corresponding to pushing the joystick shaft down. When this is triggered, we will sample the input data, decide the on/off states of each channel, and generate our own event to inform any interested classes that new data is available.

The code is fairly simple:

```
using Gadgeteer.Modules.GHIElectronics;
using Microsoft.SPOT;

namespace TestApplication.Application
{
    /// <summary>
    /// Use a joystick as a simple 2 channel input selector
    /// x axis is one channel, y the other.
    /// Joystick button is used as an action
    /// </summary>
    public class JoyInput
    {
        private Joystick m_joystick;
        private bool m_channel1On = false;
        private bool m_channel2On = false;

        private const double THRESHOLD_ON = 0.6;

        public JoyInput(Joystick control)
        {
            m_joystick = control;
            Initialize();
        }

        private void Initialize()
        {
            m_joystick.JoystickPressed += m_joystick_JoystickPressed;
        }

        // our event to notify that the input has changed
        public EventHandler OnInputChange;

        //Get channel one state
        public bool Channel1On
        {
```

```
            get { return m_channel1On; }
        }

        //Get channel 2 state
        public bool Channel2On
        {
            get { return m_channel2On; }
        }

        /// <summary>
        /// Base event handler for Joystick button press
        /// We will calculate our 2 channel states
        /// and call our input Change event
        /// </summary>
        /// <param name="sender"></param>
        /// <param name="state"></param>
        void m_joystick_JoystickPressed(Joystick sender,
                        Joystick.JoystickState state)
        {
            Joystick.Position position = m_joystick.GetJoystick
              Postion();
            double x = position.X;
            double y = position.Y;
            m_channel1On = x > THRESHOLD_ON;
            m_channel2On = y > THRESHOLD_ON;

            //Fire our notification event
            if (OnInputChange != null)
            {
                OnInputChange(this,null);
            }
        }
    }
}
```

We pass in the physical module in the constructor and then call our initialize routine. The `Initialize()` routine simply connects an event handler to the joystick button pressed event.

Our event handler routine (`m_joystick_JoystickPressed`) will get the X and Y axes positions from the data class passed in with the event. It will then compare the X and Y axes positions with the constant value `THRESHOLD_ON` (set to 0.6). If it is greater than 0.6, the channel property is set to on; if it is less, it is set to off. We then fire our custom event `OnInputChange`.

Application Data Notification Events

We are using a "standard" predefined notification event of type `EventHandler` as defined by this line:

```
// our event to notify that the input has changed
    public EventHandler OnInputChange;
```

Any external class interested in these notifications can register an event handler to this event. We will do this in our `Initialize` function in the `DemoApp` class.

Rather than use a "notification" event (`EventHandler`), we could have defined our own custom event type and passed data into the event handler. .NET has a recommended practice for defining event delegates (the prototype of our event type). It is defined as follows:

```
delegate void [delegateName](object sender,  EventArgs);
```

It is recommended that you send a reference to the originator of the event (`sender`) and data based on the class `EventArgs`. In fact, this delegate definition is used by the standard `EventHandler` event.

The delegate part of an event is just a definition or prototype for any functions used as handlers for the event. It defines their function calling type. So a handler method for an `EventHandler` event must be in this format:

```
void [FunctionName](object sender, EventArgs e)
```

In this case, `EventArgs` is the standard base class, which actually does nothing.

If we want to define our own custom event and pass some data back to the handler, we first need to define our custom delegate; then we can define our event. For instance, we want a custom event that passes back two integers as data. We want to keep to the .NET recommended practice, so we define a "custom" `EventArgs` class to hold our data. This class inherits from the `EventArgs` class and gives public read-only access to the two data integers. It would look something like this:

```
public class CustomEventArg : EventArgs
      {
          public CustomEventArg(int arg1, int arg2)
          {
              this.Data1 = Data1;
              this.Data2 = arg2;
          }

          public int Data1 { get; private set; }
          public int Data2 { get; private set; }
      }
```

To make the listing shorter, we have used the "auto-implemented properties" shortcut:

```
      public int Data1 { get; private set; }
```

This defines an integer property, `Data1`, with public read but private write. The `CustomEventArg` class gives us read access to both of our integer data values.

Now we define a delegate, a prototype for any handler methods:

```
public delegate void customEventDelegate(object sender,
CustomEventArg);
```

As you can see, we have a reference to the instance that generated the event (`sender`) and the new `CustomEventArg` class containing our data.

Now we can define the actual public event used by external classes to attach an event handler to, as shown next.

```
public event customEventDelegate OnCustomEvent;
```

There can be several handlers, one handler, or no handlers registered with our event. When we fire the event, all the handlers attached to it will be called, sequentially. When we want to fire (or invoke) the event, we first check whether it is null. If it is null, we have no handlers registered, so we don't fire it. If it is not null, handlers have been registered to it.

The most common way to fire the event is on the current operating thread. To do this, we invoke the event, as in the following:

```
//Fire our notification event
        if (OnInputChange != null)
        {
            OnInputChange(this,null);
        }
```

This can also be coded as follows:

```
OnInputChange.Invoke(this,null);
```

`OnInputChange(this,null)` is just a shortcut way of typing `OnInputChange` `.Invoke(this,null)`.

It is also possible to marshal the event onto another thread, but this is a more advanced topic.

NOTE *For a more complete discussion on events and delegates, refer to the Microsoft MSDN help documentation, or just Google "C# events."*

DemoApp Class

Our application is contained in the class **DemoApp**. This class is a container for our sensor classes and uses the data from those classes to control our output type modules. The application is event driven and responds to the `inputchanged` event from the **JoyInput** class and outputs the required response to the event to the **LedDisplay** class. The **DemoApp** class also uses a timer to poll the **AccelMod** sensor class and outputs the X and Y data to the **OledDisplay**. All the required Gadgeteer module instances are passed to the constructor. The complete code listing is shown here:

```
using AccelDemo;
using DaisyLink;
using Gadgeteer;
using Gadgeteer.Modules.GHIElectronics;
using Gadgeteer.Modules.Seeed;
using Gadgeteer.Modules.Sytech;
using Microsoft.SPOT;
using GT = Gadgeteer;

namespace TestApplication.Application
{

    public class DemoApp
```

```csharp
{
    private AccelMod m_accel;
    private LedDisplay m_ledDisplay;
    private JoyInput m_joycontrol;
    private OledDisplay m_oledDisplay;

    private bool m_led1 = false;
    private bool m_led2 = false;
    private Timer m_accelPoll;

    private Font m_font;

    public DemoApp(Accel3Axis accel,
                   MulticolorLed led1,
                   MulticolorLed led2,
                   Joystick joystick,
                    OledDisplay display)
    {
        // create our app elements
        m_accel = new AccelMod(accel);
        m_ledDisplay = new LedDisplay(led1, led2);
        m_joycontrol = new JoyInput(joystick);
        m_oledDisplay = display;
        Initialize();
    }

    /// <summary>
    /// Initialize our app
    /// </summary>
    private void Initialize()
    {
        // init our joystick switch and LEDs functionality
        m_joycontrol.OnInputChange += OnInputSelect;
        //init our accelerometer and Oled display
        m_accelPoll = new Timer(1000);
        m_accelPoll.Tick +=
              new Timer.TickEventHandler(m_accelPoll_Tick);
        InitOledDisplay();
        m_accelPoll.Start();

    }

    /// <summary>
    /// Init the Oled display
    /// </summary>
    private void InitOledDisplay()
    {
        m_oledDisplay.SimpleGraphics.Clear();
        m_oledDisplay.SimpleGraphics.BackgroundColor = Color.White;
        m_font = Resources.GetFont(Resources.FontResources.small);
        //turn off auto redraw — display update is faster this way
        m_oledDisplay.SimpleGraphics.AutoRedraw = false;
        m_oledDisplay.SimpleGraphics.DisplayText("X Axis: 0",
          m_font,Color.Black,10,10);
```

```
        m_oledDisplay.SimpleGraphics.DisplayText("Y Axis: 0",
        m_font,
Color.Black, 10, 40);
        m_oledDisplay.SimpleGraphics.Redraw();
    }

    /// <summary>
    /// Timer tick handler
    /// Executes on the main thread
    /// </summary>
    /// <param name="timer"></param>
    void m_accelPoll_Tick(Timer timer)
    {
        // get new data
        short xdata = m_accel.GetXdata();
        short ydata = m_accel.GetYdata();

        //clear the display
        m_oledDisplay.SimpleGraphics.ClearNoRedraw();
        //update the display
        m_oledDisplay.SimpleGraphics.DisplayText(" X Axis: " +
          xdata.ToString(),m_font,Color.Black,10,10);
        m_oledDisplay.SimpleGraphics.DisplayText(" Y Axis: " +
          ydata.ToString(), m_font, Color.Black, 10, 40);
        //now draw the display
        m_oledDisplay.SimpleGraphics.Redraw();
    }

    /// <summary>
    /// JoyControl input changed event
    /// Execs on main thread as base event
    /// is on main thread
    /// </summary>
    /// <param name="sender"></param>
    /// <param name="e"></param>
    private void OnInputSelect(object sender, EventArgs e)
    {
        if (m_joycontrol.Channel1On != m_led1)
        {
            m_led1 = m_joycontrol.Channel1On;
            m_ledDisplay.BlinkLed1Color(m_led1 ? GT.Color.Green :
              GT.Color.Red);
        }
        if (m_joycontrol.Channel2On != m_led2)
        {
            m_led2 = m_joycontrol.Channel2On;
            m_ledDisplay.BlinkLed2Color(m_led2 ? GT.Color.Green :
              GT.Color.Red);
        }
    }
  }
}
```

In the constructor, we use the passed-in instances of the Gadgeteer modules to create our sensor classes. We then call our **Initialize()** routine. The **Initialize()** routine connects any event handlers to the sensor classes—in this case, just the **JoyInput** event. We create our polling timer to be used to poll the **AccelMod** sensor class and initialize the **OledDisplay**.

The **OledDisplay** initialization sets the font we want to use (getting it from the program resources) and sets the screen background color. To speed up writing to the display, we turn `AutoRedraw` off. This means when we make a call to add text (or graphics) to the display, only the memory buffer is updated; the memory buffer is not actually written to the display, because we will do this manually by calling the function `Redraw`. On our display, we are updating two separate lines of text. This way, we can quickly change the buffer memory for both lines, and then when the complete screen is composed, draw the new screen in one operation.

We start off by initializing the two text lines with the X and Y values of 0.

Our event handler for the **JoyInput** event is **OnInputSelect()**. Here we get the two channel states (on or off) and change the LED color for that channel LED to red for off and green for on.

Shortcut Notation for "If ..Else" Statements

To set the LED color, we are looking at the channel state. Our test is *"If channel on, then LED is green, else LED is red."* We use a coding shortcut to achieve this—the C# `? :` operator. Consider the following syntax:

```
condition ? first-expression : second_expression
```

This means, if the *condition* is true, execute the first statement; if the *condition* is false, execute the second statement. The *condition* must be a Boolean expression.

So the following code,

```
m_ledDisplay.BlinkLed1Color(m_led1 ? GT.Color.Green :
GT.Color.Red);
```

is read as

```
if m_led1 is true, set BlinkLed1Color(Green)
else m_led1 is false, set BlinkLed1(Red)
```

Our timer event handler is called for every timer tick event in `m_accelPoll _Tick()`. The handler will get the X and Y data from the **AccelMod** sensor class. The handler will format the two text lines, showing the values, on the **OledDisplay**. But we need to clear the display screen first; otherwise, the values will be added onto the previous text, and the text will be unreadable. After we have formatted the entire display, we call the `ReDraw()` method to render (draw) the display.

Gadgeteer Program.cs

All that remains to do is to integrate our application class **DemoApp** into the Gadgeteer application framework. The **Program.cs** class was generated by the Gadgeteer project template. The designer allowed us to add the required Gadgeteer modules and initialize them. Our code to tie all the pieces together is simple and is shown next.

```
using Microsoft.SPOT;
using TestApplication.Application;

namespace TestApplication
{
    public partial class Program
    {
        // our app class
        private DemoApp testApp;

        // This method is run when the mainboard is powered up or reset.
        void ProgramStarted()
        {
            // create our event driven application
            testApp = new DemoApp(accel3Axis,led,led1,joystick,
                oledDisplay);

            // Use Debug.Print to show messages in Visual Studio's
            // "Output" window during debugging.
            Debug.Print("Program Started");
        }
    }
}
```

We create an instance of our **DemoApp** class, called `testApp`, and pass in the required instances of the Gadgeteer modules to the constructor. Because our application class is event-driven, there is nothing left to do.

Summary

In this chapter, we explored how we can separate the required sensor functions into separate, independent classes. This allows us to focus on the functionality required by the sensor class and allows us to test the classes independently. Separating out the code into classes also promotes the principle of reuse of the code in different projects.

We then designed our complete application with its own class, using our sensor module classes. This allowed us to design and code the interaction between the sensor data and the actions required as a result of the data, without being concerned about how the data is generated. Working with structured layers helps us make the code testable and makes it easier to modify the final functionality of the design.

We also looked at some basic coding methods for our sensor modules, such as threads and events.

CHAPTER 9

Serial Communications Projects

This chapter explores the use of serial communications and serial-based modules in projects. If you use generic serial modules, such as a Serial2USB or an XBee module, you'll get an instance of the Gadgeteer serial interface class to use in your application. This is a wrapper around a .NET SerialPort that includes additional functionality to make reading and writing a bit simpler. These general-purpose serial modules provide access to the serial port, allowing you, the programmer, to decide exactly how you want to use the port.

Supplying just a serial interface class is more obvious with the Serial2USB module, because it converts a serial port to a USB virtual serial port, but it's not immediately obvious with an XBee module. The XBee module is just a "carrier" board, which means it allows a number of different modules to be plugged into it. If the module is not an XBee module, you can plug in a WiFly WLAN module. All these plug-in modules are controlled by a different serial protocol and can be used in different ways and modes, so it is left to the programmer to implement the final protocol using the Gadgeteer serial interface.

Serial projects require a couple of basic elements:

- You need to be able to send and/or receive data using an asynchronous connection.

- Your connection must work at a set physical speed (baud rate) using some specific settings, such as the number of stop bits, use of hardware handshake for control or no handshake, and so on.

The Gadgeteer serial interface, combined with the Micro Framework .NET serial class, provides both of these elements for you. All you need to do is concentrate on the contents of the data, how it is formatted, and what actions need to be taken on receiving or sending the data.

Let's look at using the serial interface to implement the basic project requirements.

Building a Serial Comms Project Using a Serial2USB Module

This serial communication project will use a Serial2USB module. This module creates a USB virtual serial port that is connected to a Gadgeteer socket capable of supporting serial communications, with or without handshake, and socket types K,U (serial support with or without hardware handshaking). The module firmware creates a Gadgeteer serial interface with a default baud rate of 38,400, 8 data bits, 1 stop bit, and no parity. Initially, we will use the default serial port settings; this takes care of the physical communications settings requirement. We will then concentrate on the first requirement: sending and receiving data.

The more common data scenario involves the ability to send and receive strings or streams of data with a terminating character that marks the end of the data stream. Common uses of this scenario would be for data logging and communicating with serial devices such as Global System for Mobile Communications (GSM) modems and Global Positioning System (GPS) units.

To communicate with a GSM unit, control sequences are sent to the modem as *AT commands*, formatted text strings that start with the characters "AT" and terminate with a carriage return and optionally a line feed (characters 0x0D, 13, and 0X0A, 10). Standard GPS data is formatted into NMEA (National Marine Electronics Association) strings, which terminate with a carriage return. We will implement a simple class that is capable of sending and receiving formatted strings. We will create our project, add the Gadgeteer modules, and implement our serial handler step-by-step, adding functionality as we go.

> **NOTE** *We have discussed creating a Gadgeteer project in Visual Studio in Chapter 2 and general design of a project in Chapter 7, so we will jump straight into the project.*

Create the New Project

Create a new Gadgeteer project using the template in Visual Studio. When the designer loads up, add the mainboard, a power supply/USB device module (to allow connection to Visual Studio for debug and deploy), and a Serial2USB module. Connect the two modules to the relevant sockets on the mainboard, using the designer (see Figure 9-1). Now connect the real hardware and use the same sockets that you selected in the designer.

The designer will create the stubs for the new project, create and initialize our modules, and then connect the modules to our mainboard.

It will add Program.gadgeteer (with two partial class files, as discussed in previous chapters) and a Program.cs file. The modules are created and initialized in the Program.generated.cs file. Do not edit this file because it is maintained by the designer. Our code will be placed in the Program.cs file.

The Serial2USB module exposes a Gadgeteer Serial port; this is the serial port used by the module. We will use the serial instance in our application.

The Gadgeteer serial class is a wrapper around the .NET Micro Framework serial class and, among other things, adds some handling for sending and receiving strings. In our first project, we receive a string sent to the board, output this string to the debug channel, and return the string to the terminal application on the PC. We add "Gadgeteer RX:" to the beginning of the string. We use a simple terminal application running on a PC to test the application.

> **NOTE** *You can use a terminal application of your choice, as long as it is capable of sending text and setting the end of the string sequence to a carriage return and carriage return/new line.*

Sytech.USBDevice   Sytech.Serial2USB

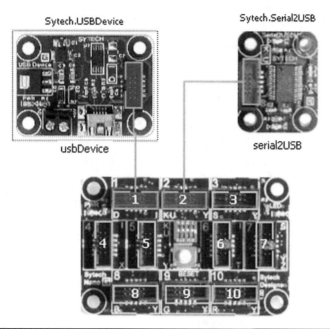

usbDevice   serial2USB

Figure 9-1 Project modules in designer

Windows HyperTerminal can be used for this, but I prefer the free terminal application from Digi for XBee development, the X-CTU. For more complex serial protocol development, X-CTU allows you to enter the serial data as a sequence of hex values. You can find X-CTU on the Digi.com web site under Support, or Google "X-CTU download."

Now let's implement our application code in its own class, as discussed in Chapter 7. Add a new class to the project and name it SerialApp.

We need an instance of the serial port to use, so pass this into our app class as a parameter in the constructor. In our constructor, save the serial port instance as a private field, so the class always has access to it. Initialize the serial port, using the default values of 38,400 for the baud, 8 data bits, and 1 stop bit. We are not using hardware handshake.

The Gadgeteer serial class adds methods to write a buffer (byte array) or string to the serial port. It also adds an auto-read function, with a terminating character sequence. For now, leave the terminating character sequence set to the default value of carriage return (**0x0D** or **\r**). C# lets us specify control characters with a \ followed by a character (**\r** indicates a carriage return).

Data being received by the serial port (from an external source) will arrive as a sequential stream of characters. The Gadgeteer serial class adds functionality that will read each character received by the serial port and collect them into a buffer. It also looks for a defined terminating sequence of characters, representing the end of a message. When it finds this terminating string, it triggers an event, **LineReceived**, passing the composed data from the buffer, minus the terminating sequence. All this is working on a background thread, so it won't block the application or Micro Framework from running. This process is normally turned off, so you need to enable it by setting the property **Serial.AutoReadLineEnabled** to **true**.

> **Important Note**
> The "Auto Read" function, in the Gadgeteer serial interface, is designed to receive 7-bit ASCII characters. Any characters with the 8th bit set (hex value greater than 0x7f) or equal to 0x00 will be discarded.

Now set the terminating character sequence to carriage return (**0x0d** – **\r**). We use a carriage return because it is the default termination for HyperTerminal and also X-CTU. When a complete string is received and terminated by a carriage return, the event is fired and the string passed—minus the carriage return. Simply attach a handler to this event, print the received string to the debug output, and then echo the string back to the sender. Add the text "Gadgeteer RX:" to the beginning of the string before you send it.

Here is the complete listing for our **SerialApp** class:

```
using Microsoft.SPOT;
using Gadgeteer.Interfaces;

namespace Serial01
{
    public class SerialApp
    {
        private Serial m_port;

        public SerialApp(Serial port)
        {
            m_port = port;
        }

        public void StartApp()
        {
            // print out port settings
            Debug.Print("CommPort:" + m_port.PortName +
                " Baud:" + m_port.BaudRate +
                " DataBits :" + m_port.DataBits +
                " Stop Bits :" + m_port.StopBits +
                " Parity :" + m_port.Parity);

            InitAutoRead();

            // now open the port -
            // use default settings
            m_port.Open();

        }

        /// <summary>
        /// Enable line auto read
```

```
/// </summary>
private void InitAutoRead()
{
    // set the terminating char to carriage return
    m_port.LineReceivedEventDelimiter = "\r";
    m_port.AutoReadLineEnabled = true;
    // now set up a line rx handler
    m_port.LineReceived += new Serial.LineReceivedEventHandler
    (m_port_LineReceived);

}

/// <summary>
/// LineReceived Event handler
/// </summary>
/// <param name="sender"></param>
/// <param name="line"></param>
void m_port_LineReceived(Serial sender, string line)
{
    Debug.Print("Line Rx : " + line);

    // now echo string back out
    string echoString = "Gadgeteer RX:" + line;
    m_port.WriteLine(echoString);
}
    }
}
```

Now modify the Program.cs class, generated by the project template, to use our new serial application in the SerialApp.cs file.

Use a Gadgeteer Timer to start the **SerialApp** after the main application starts running. You don't actually need to do this, because our application is triggered by Gadgeteer serial events. These will not be generated until after the main application is running, but this use of the timer is a useful example of how to start your application after the main framework has started.

Our code is added to the **ProgramStarted** method.

Now add a new instance of our **SerialApp** class, passing in the **Gadgeteer.Serial** instance from the Serial2USB module:

```
m_application = new SerialApp(serial2USB.GetPort);
```

Next, create a **Gadgeteer.Timer**, with a 100 millisecond delay and a handler for the tick event. The timer is set to run once. After the tick event is triggered the first time, the timer will stop. Finally, start the timer.

NOTE *Do not forget this step, or your timer tick event will never fire, and your serial application will never start!*

```
GT.Timer timer = new Timer(100,Timer.BehaviorType.RunOnce);
        timer.Tick += new Timer.TickEventHandler(timer_Tick);
        timer.Start();
```

Now add the timer tick event handler **timer_Tick**. In this handler, we call **StartApp** from our new **SerialApp** class. This will start running our communications handler, after the main application has started running.

```
void timer_Tick(Timer timer)
        {
            m_application.StartApp();
            Debug.Print("Start Serial Test Application");
        }
```

The complete code listing for Program.cs is shown next. Note that I have removed unused "using" directives from the top of the file to make the code listing slightly shorter.

```
using Microsoft.SPOT;
using GT = Gadgeteer;
using Timer = Gadgeteer.Timer;
namespace Serial01
{
    public partial class Program
    {
        private SerialApp m_application;
        // This method is run when the mainboard is powered up or reset.
        void ProgramStarted()
        {
            m_application = new SerialApp(serial2USB.GetPort);
            GT.Timer timer = new Timer(100,Timer.BehaviorType.RunOnce);
            timer.Tick += new Timer.TickEventHandler(timer_Tick);
            timer.Start();
            // Use Debug.Print to show messages in Visual Studio's
            // "Output" window during debugging.
            Debug.Print("Program Started");
        }

        void timer_Tick(Timer timer)
        {
            m_application.StartApp();
            Debug.Print("Start Serial Test Application");
        }
    }
}
```

You can check the code for syntax mistakes and typing errors by doing a quick build. In Solution Explorer (usually on the right side of the screen), right-click the project name, Serial01, and choose Build, as shown in Figure 9-2.

If all is well, the build will be successful, with no errors. You will see the build progress in the output window of Visual Studio; here's the last line:

```
"======= Build: 1 succeeded or up-to-date, 0 failed, 0 skipped ======="
```

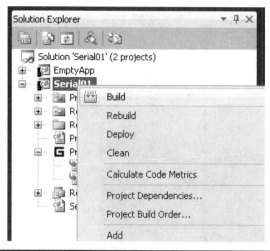

FIGURE 9-2 Build the project

Start and Debug the Application

Now we are ready to deploy and debug the new application to our hardware. Make sure the modules are plugged into the correct sockets on the mainboard.

1. Connect up your mainboard to your PC using a USB cable plugged into the USBDevice module.

2. Connect up a second USB cable to the Serial2USB module and your PC. This is the serial connecting to the terminal application running on the PC.

3. The mainboard will now connect the Micro Framework USB debug channel to your PC. Check to make sure that it has connected correctly by using MFDeploy, as discussed in Chapter 6.

4. Start your terminal application on the PC and connect it to the virtual comm port created by the Serial2USB module.

5. Right-click the project name in Solution Explorer and choose Debug | Start New Instance (see Figure 9-3).

   This will build and then deploy your new application to your mainboard. After the application is deployed, the debugger will be connected and the application will start running.

The progress of this process is displayed in the output window. After the build, the application will be deployed to the device. After the code is deployed to the device (this may take a few seconds), a report on all the DLL assemblies (sections of code) being used in the application will be generated. These DLLs will mainly be the Micro Framework and Gadgeteer libraries. After all the assembly reports, the debugger will start loading the application.

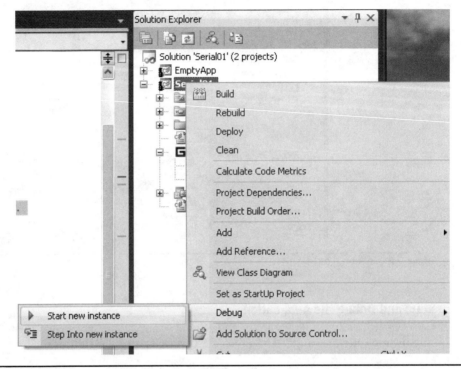

Figure 9-3 Debug and start new instance

You'll see a list of all the assemblies loaded, and the actual application will be started. At this point, you'll see any debug text you added to the application. The startup of our serial application includes some debug text, which looks like this:

```
The thread '<No Name>' (0x2) has exited with code 0 (0x0).
Using mainboard Sytech Designs Ltd Nano version 1.0
Program Started
CommPort:COM1 Baud:38400 DataBits :8 Stop Bits :1 Parity :0
Start Serial Test Application
```

Our application is running and the Visual Studio debugger is connected. Now we need to start our terminal application running on the PC using the virtual serial port created by the Serial2USB module.

Start the Terminal Application

Start the terminal program of your choice, and connect to the virtual serial port, used by the Serial2USB module. If you are using X-CTU as the terminal program, the start screen will list all the serial ports available for use. The one we need is usually labeled "USB Serial Port (COMx)." If you are not sure which one it is, follow these steps:

1. Disconnect the USB cable from the Serial2USB module.

2. Start up X-CTU and note the serial ports listed.

3. Close X-CTU, reconnect the USB cable to the Serial2USB module, and restart X-CTU. The new serial port that appears on the list is the one you need.

You can also use Windows Device Manager to identify your USB serial port.

1. Right-click Computer (My Computer for the XP guys), select Properties, and then select Device Manager (Hardware | Device Manager on XP).

2. Click the Ports (COM & LPT) section to expand it. All the comm ports will be listed here (see Figure 9-4).

3. If you are not sure which is the correct one, unplug the USB cable, note which ports are available, plug the USB cable back in, and look for the added entry; this is your comm port. Note that it takes a second or two for the USB device to register and unregister when connecting and disconnecting from the PC.

The PC terminal program needs to be started, using the comm port connected to our hardware, with the initial settings of 38400 baud, no flow control, 8 data bits, no parity, and 1 stop bit. For X-CTU, this looks like Figure 9-5.

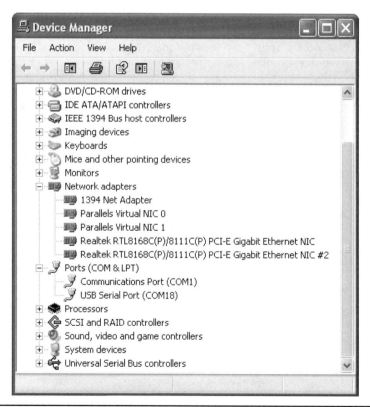

FIGURE 9-4 Device Manager USB comm port number

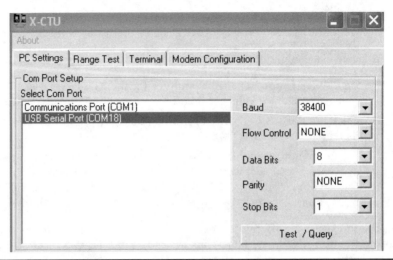

FIGURE 9-5 X-CTU PC Settings screen

Now do the following:

1. Select the Terminal tab to bring up the terminal screen. The terminal program should be running and connected to the comm port. I am using X-CTU because of the extra debugging and control features it provides. Technically, it is designed to talk to Digi XBee modules and allows direct entry and viewing of the protocol commands. These features make it useful as a tool in comms projects.

2. Turn on the Hex feature so you can see the ASCII text on the screen and the corresponding hex values. In the hex window, you'll also see any hidden characters such as carriage returns, and so on. Click the Show Hex button to the right of the menu bar.

3. The output screen is now split. The section on the right is hex, and the section on the left is ASCII. Place the cursor on the left part of the screen and type **this is a test**, and end the line with a return (**0x0D**). This will send the characters to the hardware, using the serial port. When the device sees the carriage return (0x0D), it will fire the line-received event handler. This code will take the string received (minus the terminating character) and prefix "Gadgeteer RX:" to the string and send it back to the PC terminal program. It will also output the string received to the Visual Studio output window as debug text:

```
void m_port_LineReceived(Serial sender, string line)
    {
        Debug.Print("Line Rx : " + line);

        // now echo string back out
        string echoString = "Gadgeteer RX:" + line;
        m_port.WriteLine(echoString);
    }
```

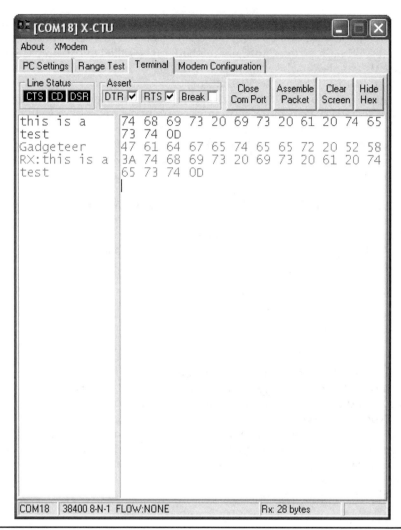

FIGURE 9-6 Test string sent and returned

If all is well, you will see the string sent from the device in the window of the PC terminal program. Figure 9-6 shows the result. You can see the ASCII text on the left and the actual hex values on the right. Note the addition of the terminating character carriage return (**0xD**), added by the function **WriteLine**. If we had used **m_port.Write()**, no terminating characters would have been appended to our string, so a carriage return would not have been sent.

Changing Serial Port Physical Settings

Our Serial2USB module serial port is created with the default physical settings of 38400 baud, 1 stop, 8 data, and no parity. If we want to change one or more of these physical settings, we can. For instance, suppose our baud rate needs to be 115200, not 38400. We can make the change in one of two ways.

We can call the **Configure** method on the Serial2USB instance and pass in all the settings at once. However, we have to do this before the serial port is created by the Serial2USB class, so the **Configure** method can be called only before **GetPort** is called. This also applies to the similar GHI versions of serial modules, such as **UsbSerial**, except their **GetPort** is called **SerialLine**.

After the serial port has been created, we can still change physical properties. Each of the settings is available as an individual property, so we can use the **Baud** property to change the baud rate. You cannot change physical properties on an open comm port. You need to call **[port].close** first, change the properties, and then you can reopen the port. You can check whether a port is open by calling the **IsOpen** property. This will return **true** if the port is open.

To demonstrate these methods, we'll modify our simple serial project to start the comm port at the 115200 baud rate. Then, when we start the comm application, we'll change the baud rate back to 38400. This will demonstrate the two mechanisms for changing physical settings: upon creation and while running.

In the Program.cs file, add the following line before the line that creates our serial application:

```
//Set serial port to 115200 baud
        serial2USB.Configure(115200, Serial.SerialParity.None,
        Serial.SerialStopBits.One, 8, false);

        m_application = new SerialApp(serial2USB.GetPort);
```

We are calling the **serial2USB.Configure** method before we make a call to **GetPort**, so the serial port has not been created at this point. This sets the port to 115200 as the initial baud rate, instead of the default 38400 setting.

Now we'll modify the **SerialApp** class to show that the port is working at 115200 and then we will change the baud rate back to 38400. In the SerialApp.cs file, modify the **StartApp** method. After we open the port, we will send a test string using the current baud rate (115200). Then we will close the port and change the baud rate back to 38400 and then open the port again.

Here is the modified code:

```
public void StartApp()
        {
                // print out port settings
                Debug.Print("CommPort:" + m_port.PortName +
                    " Baud:" + m_port.BaudRate +
                    " DataBits :" + m_port.DataBits +
                    " Stop Bits :" + m_port.StopBits +
                    " Parity :" + m_port.Parity);

                InitAutoRead();

                // now open the port -
                m_port.Open();

                // send a test string at 115200
                m_port.WriteLine(" Start at 115200 baud - now changing to
                38400");
```

```
        // you cannot change physical settings on an open comm port
        m_port.Close();
        m_port.BaudRate = 38400;
        Debug.Print("change to 38400");
        // don't forget to reopen the port
        m_port.Open();
    }
```

To test this, do the following:

1. Open your terminal application and set the comm port settings to 115200.
   Figure 9-7 shows X-CTU with the baud set to 115200.

2. Start the terminal application on your PC.

3. Build and deploy the modified serial application. After the application is loaded
   and running, the comm port is now initially set to 115200 baud and the test
   string "Start at 115200 baud – now changing to 38400" is sent at this speed and
   will be displayed on the terminal application (see Figure 9-8).

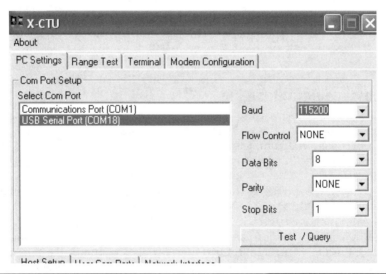

FIGURE 9-7 Terminal set to 115200

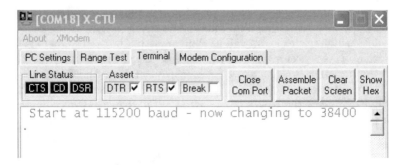

FIGURE 9-8 Test string at 115200

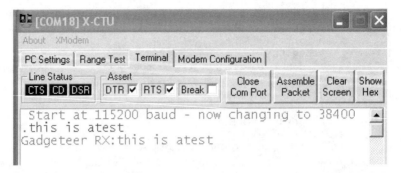

FIGURE 9-9 Test string working at 38400

4. Now go back to the PC Settings tab in X-CTU, and change the baud rate to 38400.

5. Return to the terminal screen, and X-CTU will close and open the port so the new settings take effect. Now the terminal is connected at 38400.

6. Type in a test string and carriage return and the device will echo the string at the baud rate of 38400 (see Figure 9-9).

Serial Message Data Handling

The project code discussed so far can be used if you simply need to send and receive strings; however, it is more likely that you'll need to implement a serial message protocol, where the complete messages are encoded data buffers. Let's look at an example of how to handle a serial protocol.

In this example, we will look at decoding NMEA-encoded sentences from a serial GPS receiver. We will decode one of the message types (several different message types are defined in NMEA): the GGA type message that contains fix data.

NMEA messages start with the string **$GP**, followed by three characters identifying the message type. This is followed by the message data and the message is terminated with **<CR><LF>** (**Carriage Return, Line Feed**). The number of data bytes and their meanings are defined by the three-character message type. Because this is an ASCII-based message protocol, we can use our Gadgeteer serial auto-read event, defining the terminating characters as **<CR><LF>** (using the c# notation "**\r\n**").

Each message data field is separated by a comma (comma-delimited). Messages can have an optional checksum; if the checksum is used then an asterisk (*) follows the last data field, with the checksum value following that. The final element is the terminating characters. The checksum is the "exclusive or" of all the characters between the **$** and the *****. Convert the value to a two-character string and that should be the same as the two checksum characters. We will not be validating the checksum in our example. This is a simple example of the basics of decoding an NMEA message that demonstrates the basic principles to get you started.

The GGA sentence is defined as follows:

```
FixTime,Latitude,N,Longitude,E,FixQuality,Number of
Satellites,HDOP,Altitude,Sea Level,empty,empty,checksum
```

We can extend out serial app class to partially decode this NMEA string. First, we change the terminating character string to **<CR><LF>** (using the c# notation "**\r\n**"). In the SerialApp.cs file, modify the method **InitAutoRead** as follows:

```
private void InitAutoRead()
        {
            // set the terminating char to carriage return, line feed
            m_port.LineReceivedEventDelimiter = "\r\n";
            m_port.AutoReadLineEnabled = true;
            // now set up a line rx handler
            m_port.LineReceived += new
Serial.LineReceivedEventHandler(m_port_LineReceived);
        }
```

Now we add our NMEA sentence decoder method. Add the following method to the code:

```
 private void HandleRxMessage(string strMessage)
        {
            if (strMessage.Substring(0,6).Equals("$GPGGA"))
            {
                // we have a fix string, split comma seperated values
                string[] gpsData = strMessage.Split(new char[] {','});
                string longStr = gpsData[2] + gpsData[3];
                string latStr = gpsData[4] + gpsData[5];
                string fixQuality = gpsData[6];
                string numbOfSatsStr = gpsData[7];
                string HDOP = gpsData[8];

                Debug.Print("GPS:" + longStr + ":"
                    + latStr + ":" +
                    "fix :" + fixQuality +
                    " Satellites " + numbOfSatsStr +
                    " HDOP :" + HDOP);

                m_port.WriteLine("Longitude :" + longStr);
                m_port.WriteLine(("Latitude :" + latStr));
            }
        }
```

We will pass any strings received in our **LineReceived** handler to this method. Modify the **m_port_LineReceived** handler method as follows:

```
void m_port_LineReceived(Serial sender, string line)
        {
            Debug.Print("Line Rx : " + line);
            HandleRxMessage(line);
        }
```

We can check the start of the string to see if it is an NMEA GGA message by examining the first six characters to see if they are **$GPGGA**. If they are, we will split the NMEA string into separate strings. Remember that NMEA data fields are separated by commas. We can use the **string.split** method to do this. We set the separator characters, in this case just a comma, and the function will split the string into an array of strings.

Next, we load the relevant message field from the string array into our data string.

So the longitude field is the third and fourth data fields. We decode only some of the fields. Next, we output the values as debug text, and finally we send the longitude and latitude back out of the serial port.

We use X-CTU to simulate the data from the GPS receiver by entering our NMEA sentence into X-CTU and sending it. We will see the decoded longitude and latitude sent from our device on the terminal screen.

Here's our test NMEA string:

```
$GPGGA,123519,4807.038,N,01131.000,E,1,08,0.9,545.4,M,46.9,M,,*47 <cr><lf>
```

X-CTU has an "Assemble Packet" function that will bring up a new window, where we can enter our complete message, as either hex or ASCII. We can assemble our complete message and then send it out of the serial port as a complete message. The easiest way is to type the message in ASCII, and then switch to hex and add the **0x0d,0x0a** for the terminating sequence <cr><lf> (carriage return, line feed).

Figures 9-10 and 9-11 show the assemble window with the message in ASCII and in hex.

Now build and deploy the new serial application and start the debugger. Enter the NMEA string into the Assemble Packet window and click Send Data. The message will be decoded and the longitude and latitude will be sent back to the terminal program. Figure 9-12 shows the result.

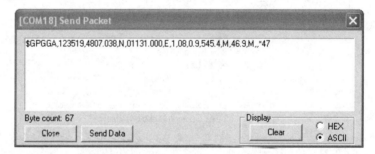

FIGURE 9-10 Assemble Packet ASCII data

FIGURE 9-11 Assemble Packet hex data

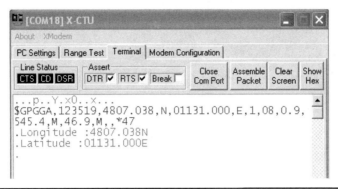

FIGURE 9-12   Complete data transfer

We have covered the general principles of implementing ASCII-based message protocols. There are many examples of serial ASCII-based protocols. Two of the most common are AT commands for talking to modem/GSM type devices and NMEA sentences from GPS receivers. With the basics covered in this chapter, you should be able to tackle these with ease. The subject of serial protocol handling could fill an entire book on its own, but we will have to leave it here.

CHAPTER 10

SD Card and File Projects

In this chapter will we look at reading and writing data files and the use of the file system on data storage devices. Gadgeteer supplies a utility class (**StorageDevice**) to assist in this. SD Gadgeteer modules should make an instance of this class available, attached to the currently inserted SD card. The **StorageDevice** class allows you to do the following:

- List directories
- Create directories
- List files in a directory
- Create/open files for writing
- Open files for reading
- Load an image from a file (jpeg, bmp, or gif)
- Delete a directory

The **StorageDevice** class uses standard .NET file I/O functions to perform these tasks. The read and write functions use binary data (arrays of bytes). If we want to read or write a text file, we either have to convert text to/from binary or use .NET stream files.

Strangely, there is no delete file function in the **StorageDevice** class. We need to use .NET file functions directly to fill in the gaps. We will look at how to use these .NET functions in this chapter. This chapter also includes projects to persist (save) application data to binary files to make data from an application session available for other sessions after power cycles. We will also cover projects to save text data in a logging function, and text-encoded record-based data saved as CSV (comma-separated values) for importing and exporting tabular data from spreadsheets and databases.

Mounting and Unmounting Removable Media

Storage devices are referred to as "volumes" in the Micro Framework and treated like a directory (unlike on a PC, where the root element of a drive is a letter, such as C:). A storage device will have a root directory name to identify it, and all operations are relative to this root name.

An SD card is removable media that may or may not be inserted. When a card is inserted, it is mounted to the file system. Normally in a Micro Framework port, when a card is inserted it is detected and automatically mounted. Then a Micro Framework event is fired, notifying the managed code of the event; a similar event is fired when the card is removed.

NOTE *Some mainboard hardware does not automatically mount detected SD cards (GHI, for instance), so you need to mount the card manually. However, beginning with version 4.1.0.500 of the Gadgeteer framework, hooks have been added to allow the Gadgeteer framework to mount the card when detected, even on devices that don't normally support this function in their Micro Framework port. So beginning with Gadgeteer version 4.1.0.500, you do not need to worry about mounting removable media manually.*

Once the card is mounted, the manufacturer-supplied Gadgeteer device firmware provides a Gadgeteer storage device instance. This class is a Gadgeteer interface and is hardware independent.

We will first look at how we obtain the storage device instance from two different mainboards—the Sytech NANO and the GHI family of devices. The difference is in the API for informing the application that an SD card has been inserted.

GHI mainboards define two events:

- **SDCardMounted(***sender, StorageDevice***)**
- **SDCardUnmounted(***sender***)**

The Sytech NANO SD mainboard has one combined event:

- **OnMediaChanged(***sender, StorageDevice, isCardInserted***)**

The end result is the same with either mainboard: when a card is inserted, you are informed, and the **StorageDevice** instance is passed to you. When a card is removed, you are notified. Once a card has been inserted and the volume is mounted, applications can have common code to handle reading and writing files, independent of the hardware platform used.

Our example application class will be written to be independent of the notification mechanism. It will provide **CardMounted(***StorageDevice***)** and **CardUnmounted()** methods. Internally, it will use the Gadgeteer **StorageDevice** and .NET file IO (input/output) methods.

The relevant SD card module mounted/unmounted event handlers will call the application class mounted/unmounted methods, passing in the **StorageDevice** class. This way, we can focus on SD file operations, independently of the hardware we are using.

GHI-Based Mainboard

Let's work with a GHI-based mainboard.

Create a new Gadgeteer application project. Add a GHI mainboard and a GHI SD Card module. Connect the SD Card module to a valid socket on the mainboard using the GUI designer (on a Hydra board, this is socket 8).

Open the Program.cs file and add the following code to connect the mount and unmount events.

SD Card Physical Interfaces

SD cards support several physical interfaces to a controller. The two most common are the 4-bit interface and the Serial Peripheral Interface (SPI). In general, an SPI usually does not support the SDHC (High Capacity) interface, so it can access only cards with a capacity of 2GB or less. The 4-bit interface is faster than SPI, but not all hardware has the required SD card controller hardware. Most Micro Framework ports use SPI, because it gives a "common denominator" for the porting code, as most (if not all) ARM-based processors will support SPI, but not all have an SD card controller. There are also licensing issues with using the 4-bit interface, whereas the SPI interface is in the public domain and no fees apply. A Gadgeteer SD Card module will usually supply the SD card connector and connect the physical card connections to the mainboard; the actual implementation of the SD functions are on the mainboard, so they are mainboard-specific. In short, not all SD modules will work in all mainboards.

```
public partial class Program
    {
        // This method is run when the mainboard is powered up or reset.
        void ProgramStarted()
        {
            sdCard.SDCardMounted +=
                new SDCard.SDCardMountedEventHandler
                    (sdCard_SDCardMounted);
            sdCard.SDCardUnmounted +=
                new SDCard.SDCardUnmountedEventHandler
                    (sdCard_SDCardUnmounted);

            Debug.Print("Program Started");
        }

        /// <summary>
        /// The card mounted event handler.
        /// Call whenever an SD card is inserted
        /// </summary>
        void sdCard_SDCardUnmounted(SDCard sender)
        {
            Debug.Print("Card UnMounted");
        }

        /// <summary>
        /// The card unmounted event
        /// Called whenever a card is ejected
        /// </summary>
        void sdCard_SDCardMounted(SDCard sender, GT.StorageDevice
            SDCard)
        {
            Debug.Print("Card Mounted");
        }
    }
```

The GHI SD module provides two events, **SDCardMounted** and **SDCardUnmounted**. In **ProgramStarted**, we connect event handlers to each of these events. The **CardMounted** event handler will be called whenever the mainboard detects that a card has been inserted and it has been able to mount the volume successfully. The instance of the Gadgeteer **StorageDevice** interface is passed to the handler. The **CardUnmounted** event handler will be called every time a card is ejected. For now, our event handlers will print some debug text. We will pass the **StorageDevice** instance into our application later—at this point, we are just demonstrating how we connect to the card with GHI-based hardware.

Sytech NANO Mainboard

The NANO mainboard SD module also provides an Ethernet interface. We will not be using the Ethernet interface in these examples, however; we'll use only the SD card functions. The module SD Card firmware provides a single event for card mounting and unmounting, with a Boolean value setting if the event has been called by a card being inserted or ejected.

Create a new Gadgeteer application project. Add a NANO mainboard and a EthernetSD module. Use the GUI designer to connect the module to the mainboard socket 7.

Open the Program.cs file and add the following code to connect the mount and unmount events.

```
// This method is run when the mainboard is powered up or reset.
    void ProgramStarted()
    {
        ethernetSD.OnMediaChanged += (ethernetSD_OnMediaChanged);

        Debug.Print("Program Started");
    }

    void ethernetSD_OnMediaChanged(object sender,
                            GT.StorageDevice sdCard,
                            bool cardInserted)
    {
        if (cardInserted)
        {
            Debug.Print("Card Inserted");
        }
        else
        {
            Debug.Print("Card Removed");
        }
    }
```

We have a single event on the SD module for mount and unmounts. We connect an event handler to this event. In the event handler, we test the Boolean parameter **cardInserted**. If it is true, a card has been inserted; if it is false, a card has been ejected. When a card has been inserted, the Gadgeteer **StorageDevice** instance is supplied. When a card has been ejected, the instance of **StorageDevice** will be null (empty).

Directory and File Handling

Directory and file handling functions are supplied by the .NET System.IO library functions. These are a subset of the normal desktop .NET functions.

Volumes and Root Directories

.NET disk volumes are identified by a letter, such as C:. If you connect an SD card reader with an SD card inserted, this will show up in Explorer as a drive letter and is treated like any other drive on your PC. In the Micro Framework, storage devices are identified by a root directory name, such as \SD1. This name can vary depending on how the mainboard manufacturer implements the file system in its Micro Framework port.

The system can supply a list of installed volumes, and the root name of a volume can be read from the volume information class. All directory and file operations have to be in the root name directory or in a subdirectory below the root name. You need to supply absolute paths from the root or set the current directory property to the root directory or a directory below the root. If the root name of the SD card is SD1, you can create directories and files in this directory and its subdirectories, but you cannot access files and directories in the absolute root (\). Functions in the Gadgeteer **StorageDevice** class hide this from you by automatically adding the root directory name to paths you supply and removing the root name from functions that return a path. The root name is available from the **StorageDevice** property **RootName**. When you use the .NET I/O functions directly, do not forget about the root directory name in your paths.

Using the StorageDevice Class

The SD card has a file system–type root directory. All directories and file operations start from this root directory. The name of the root directory is supplied by the **RootDirectory** property. All directory and file functions provided by the **StorageDevice** hide this root directory—that is, when you pass a path into a function, it does not start with the root directory name but the path is appended by the **StorageDevice** class function. Likewise, any function that returns a path has the root directory name stripped off before passing the path back to the application. For example, if your device has a root directory of \SD and it has two subdirectories, Dir1 and Dir2, a call to **ListRootDirectorySubdirectories** will return a string array with two paths in Dir1 and Dir2—whereas, in fact, the directory names are really \SD\Dir1 and \SD\Dir2. (See the upcoming Tip for an explanation of the \\ notation.)

If you want to add a new directory under Dir1 called Sub1, then the path passed into the function **CreateDirectory(string path)** will be **Dir1\\Sub1 (CreateDirectory("Dir1\\ Sub1"))**. The root directory is not passed in your path; this will be added by the **StorageDevice.CreateDirectory** function. If you access a .NET directory or file function directly, do not forget to include the root directory in your path.

TIP *The separator character used in paths is \. However, in C#, this is a special character used in strings, which means that the next character is a control code. So to include the character \ in a string, we need to type \\. An alternative method of writing C# string values that ignores the control character convention is to prepend the string with the character @. This turns off the control character feature. So instead of \\Dir\\SubDir\\ File, we'd write @"\Dir\SubDir\File".*

Figure 10-1 shows the class diagram of the **StorageDevice** class. It provides properties for the **RootDirectory** and also the **Volume** for the inserted SD card. The **Volume** contains data about the SD card, such as file system (Fat16, Fat32) and the size of the card, along with the free space available for use.

Directories

The class methods provide functions to list the directories on the device. You can list all the directories in the root directory using **ListRootDirectorySubDirectories()** or provide a path to the subdirectory you want to use as a base with **ListDirectories(*string:path*)**. You can also create a new subdirectory with **CreateDirectory(*string:path*)**.

The path you pass is relative to the root directory. Do not put the root directory in your path because it is added automatically by the method. For instance, if you want a new directory [*root*]\mySubDir\mysubSubDir, you would pass **@"mySubDir\ mysubSubDir "** as the path string (see the previous Tip for an explanation of the @ notation). The function **Delete(string:path)** is used to delete a directory.

NOTE *You cannot delete a file with this function, as suggested in the IntelliSense comments, but if you delete a directory, all files in the directory will be deleted. Also, the directory must exist; otherwise a system exception will be raised, which, if unhandled, will crash your application. We will see how to trap and handle this exception in the next section, "Files."*

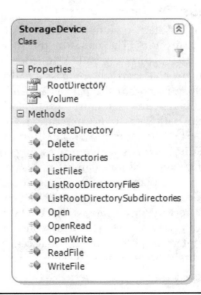

FIGURE 10-1 **StorageDevice** class diagram

Files

The **StorageDevice** class provides functions to list files, either in the root directory or in a subdirectory (**ListRootDirectoryFiles()** and **ListFiles(***string: path***)**). When using **ListFiles**, the path you provide is relative to the root directory—but you should not add the root directory name to your path. Both functions will return a string array, with an entry for each file found.

The class provides several functions to read and write files but no function to delete a file directly. We will cover how to do this at the end of this section.

The functions for reading and writing files all use a .NET **FileStream** class, which deals with binary data. If we want to deal with text files, we need to convert our text data to/from binary; however, there are better ways to do this. We will add more file functionality in this chapter's example projects, which will allow us to deal directly with text-based files such as comma-separated text files.

The functions **ReadFile** and **WriteFile** allow you to read a binary file, return the binary contents as a byte array, and write a byte array to a file. On writing, if the file does not exist, it is created; if it does exist, it is overwritten. On **ReadFile**, you pass in the path to the file. The path and file must exist; if not, a system exception is thrown, and if not handled, your application will terminate. Likewise, on **WriteFile**, the directory part of the path must exist; the file will be created if it does not exist, but if the directory part of the path for the file does not exist, then a system exception is raised; if not handled, your application will terminate. These general rules apply to all file type operations: if you try and read a file that does not exist or write to directory locations that do not exist, an exception is raised, causing your application to terminate.

You can handle this in several ways:

- Stop it happening in the first place by checking whether the path is valid.
- Catch and handle the exceptions.
- Do both of the above.

We can check if a directory or file exists with a call to the .NET System.IO **Directory. Exists(***path***)** or **File.Exists(***path***)** functions; both return a Boolean value **true** if the directory or file exists and **false** if not. If you need to check the directory path for a **WriteFile** operation, just call the **Directory.Exists** function first, passing the directory path to where you want to create the file. If it returns **true**, you can proceed with a call to **WriteFile**. If you need to read a file, call **File.Exists(***path***)** first, passing in the pathname and filename. If it returns **true**, you know it's safe to read the file.

If a file or directory function causes an error, such as when the directory does not exist, a system exception is raised. If this error is not caught and handled, it will terminate (crash) your application. We can catch any exceptions by wrapping our file-handling calls with a try-catch function. This tells the code to try and execute the following, but if there is an error, then catch it and let you handle it. If the error is caught and handled, it stops there and the application is allowed to continue.

The following code snippet shows how to do this:

```
try
{
sd.WriteFile("rubbish\\testfile.bin", buffer);
}
 catch (Exception exc)
```

```
{
Debug.Print(exc.Message);
}
```

In this code, we are trying to write a byte array buffer, **buffer**, to a file, testfile.bin, but the directory in the path rubbish does not exist. The call will generate a "directory not found" error (exception). But we enclose the call in a try statement and supply a catch statement to handle any errors—in this case, just a debug statement. Now, when we execute this **WriteFile** function, it will still fail, because we can't write to a directory that does not exist, but the error (exception) is caught and handled by our catch statement. Because we have caught and handled the error, it stops here and will not terminate our application—but remember that the file has still not been created, so your code will need to handle this.

A number of file read and write functions are included in the **StorageDevice** class. The simplest are **ReadFile** and **WriteFile**. These read or write an array of bytes to and from a file.

The other read and write functions open a file stream connected to a file; you then use the **FileStream** class to read and write your data.

NOTE *For more details on using streams and file streams in particular, refer to the Microsoft MSDN online help system. Google (or use your preferred search engine) "common I/O tasks MSDN."*

The **StorageDevice** functions **OpenRead(***string filename***)** and **OpenWrite(***string filename***)** will open a new file stream associated with the file you pass in the *filename* parameter. The write will create a new file if it does not already exist or overwrite an existing file. The read opens an existing file. Both will throw a system exception if the filename is invalid for some reason.

File streams are based on the stream class. A stream is basically a view of a sequence of bytes. It has methods that allow you to read and write data and also a pointer to mark the current position in the sequence. The file stream is a view into a file, allowing you to read and write data to it.

A file stream is a resource; when you have finished with it, you must close it by calling the **[***filestream***].Close()** function. If you do not close a file (especially one opened as a write file), then if the application terminates or the SD card is removed, it is possible to corrupt the file.

The following code sample is an example of how to use the **OpenRead** and **OpenWrite StorageDevice** functions.

```
// example of openread/openwrite usage
// sd is our StorageDevice instance
string dirName = "testdir";
string testfile = "\\testfile.bin";
sd.CreateDirectory(dirName);
FileStream fileStm = null;
byte testByte = 1;
try
{
        fileStm = sd.OpenWrite(dirName + testfile);
        while (testByte < 100)
```

```
        {
            fileStm.WriteByte(testByte++);
        }
        fileStm.Close(); // close the file
        fileStm.Dispose(); // get rid of original filestream
}
        catch (Exception exc)
{
Debug.Print("error creating file");
}
// now read test file
byte[] data;
try
{
        fileStm = sd.OpenRead(dirName + testfile);
        using (fileStm)
        {
            data = new byte[fileStm.Length];
            fileStm.Read(data, 0, (int)fileStm.Length);
            Debug.Print("read file length is " + data.Length);
        }
}
catch (Exception exc)
{
        Debug.Print("Error reading file");
}
```

First we create a new file and load it with binary values from 1 to 99. We use the **OpenWrite** function to create a new file connected to a file stream. We then use the .NET **filestream.WriteByte** function to add our binary data to the file. Then we close the file.

We wrap the open and write operations in a try-catch loop to ensure that any errors creating exceptions are handled. Next, we open our new file and read the data back. We use the **OpenRead** function to return a file stream, connect to the file, and then use the .NET **filestream.Read** function to read the data back into a byte array. We have again wrapped the open file and reading code section in a try-catch loop to protect against any file I/O errors. But we have also used a .NET shortcut and wrapped the file handling section in a using statement. The using section will protect our file stream resource and will close and dispose of the resource automatically when we have finished with it. It effectively does the same as the write section of code, but saves us explicitly closing and disposing of the file resource.

The remaining **StorageDevice** file function is **Open(***filepath, FileMode, FileAccess***)**. This is the most flexible function and allows you to open a file stream to a file, set the file mode (open, create, append, and so on) and the access mode (read, write, read and write). The following code snippets show how to open the same files used in the **OpenRead** and **OpenWrite** examples, to allow writing to the test file and reading the test file.

Open the test file for writing (create or open existing):

```
fileStm = sd.Open(dirName + testfile, FileMode.OpenOrCreate,
    FileAccess.Write);
```

Open the test file for reading:

```
fileStm = sd.Open(dirName + testfile, FileMode.Open, FileAccess.Read);
```

The final method in the **StorageDevice** allows you to read a JPEG or BMP graphics file from the SD card and convert it into a .NET **BitMap** instance, which you can use for graphics. The method is **LoadBitmap**. In the parameters you pass the file path and an enumeration defining the type of image in the file (bmp, jpeg, and so on).

The missing function from the **StorageDevice** is file deletion. We will use the **System.IO.File** class from the .NET libraries to delete files. The class is a static class, so you do not need to create an instance of it; just use the function directly. The **File. Delete(*string pathname*)** is a fairly safe method to use—if the file in the *pathname* does not exist, it does *not* throw any exceptions; if it does exist, it is deleted. The following code snippet shows how to use it to delete the test file we created in our earlier examples:

```
File.Delete(dirName + testfile);
```

TIP *For a quick reference to the Micro Framework API functions, from Version 4.2 of the Micro Framework SDK, a chm type help file is installed in the document directory of the SDK. It is located at [Program Files]\Microsoft .NET Micro Framework\v4.2\Documentation\Net MicroFramework Docs\PSDK.chm. Unfortunately, the 4.1 SDK did not supply a compiled chm file. It is recommended that you use MF4.2 and the latest Gadgeteer library code.*

That covers the basic theory on reading and writing files and directories using the **StorageDevice** Gadgeteer class. As you have seen, the class handles reading and writing binary files, plus reading a graphics file.

We will now look at some file projects to handle some real-world examples. We will look at how to persist the data fields of a class to a binary file and how to load the data values of a class from a binary file (serialize and deserialize) and also using text-based files rather than binary files. We will look at how to use text files to save data in plain text and also a comma-separated format, suitable for importing into an Excel-type spreadsheet, useful for data logging applications and importing large amounts of data into your application.

Save and Restore Setup Data Project

In this project you'll learn how to write a class with fields of setup and configuration data. We will make the class *serializable*. This means we can save and load the data contents of the class to a byte array buffer. We will save the serialized data buffer to a binary file. We can then read the buffer from the file and reload our class using the buffer. To make the class serializable, all we need to do is mark it with a special attribute label, and the .NET Micro Framework will do the rest. When we make a class serializable, all public, private, and protected fields (global variables) will be used in the serialize data. If there are fields we don't want to save, they can be marked with an attribute label so they are not included. (For a more detailed description of serialization, refer to the MSDN .NET documentation.)

We'll define example data fields to be used by the application in the class, serialize the data, and then save this to a file. Next time the application starts, if it finds the binary file with our data in it, the application will load an instance of our setup class, using the values from the binary file. This allows us to generate application data and then save it to an SD card, where it can be reused when the application next runs. This is useful for setup and configuration data that needs to be saved between application sessions.

NOTE *Micro Framework serialization produces binary data that's different from that of the equivalent desktop .NET functionality. This is because the Micro Framework version optimizes the data in a different way, so it can be stored in a more efficient manner. You cannot directly take a Micro Framework serialized data file and deserialize it into the same class on a desktop system. If you need to do this, you need to write a custom serialize-based class to handle the different data formatting.*

Adding Classes to the Project

Use the project you created to demonstrate using the media inserted/ejected events, or create a new project with a mainboard and SD card module. Add a new class called **SerializeTest** to the project for our test application. Add a class to the project called **SetupData**; this will be our serialized data class

The binary setup data file will be stored on the SD card; therefore, you cannot access it until an SD card is inserted. When you have access to an SD card, you will pass the **StorageDevice** instance into your test application **SerializeTest**. You do this by connecting an event handler to the mainboard card inserted event. As discussed in the first project, the event can be different for each type of mainboard. When the card is inserted, the **StorageDevice** instance is passed to the test application.

If the mainboard powers up with a card already inserted, you most likely will not see a card inserted event. So on power up, you need to check to see if a card is already inserted. Most manufacturers supply a **CardInserted** property in their firmware to check for a card. If a card is inserted, you use a timer to delay running the test application, so the Framework is running first. If there is no card, the card inserted event will run the test application (which will occur after the Framework is running).

You have added the test application class and the **SetupData** class to our project. Now let's add the implementation.

Implement the SetupData Class

This class has the configuration data as fields. It is marked as serializable so the fields can be converted to a byte array and loaded from a byte array. It contains some example test fields. The serialization process will use all fields—private, public, and protected. The test fields include an example string field and some integer values, plus an array of bytes.

Enter the following code into the **SetupData** class file:

```
using System;

namespace SerializeSD
{
    /// <summary>
    /// Example config data class that can be
    /// serialized
    /// </summary>
```

```
[Serializable]
public class SetupData
{
    public int SetupInt;
    public string SetupString;
    public uint NumberOfUses;
    private DateTime LastUse;
    public byte[] TestArray;

    public SetupData(string testStr)
    {
        SetupString = testStr;
        TestArray = new byte[1];
        SetupInt = 0;
        NumberOfUses = 1;
        LastUse = DateTime.Now;
    }
    public void IncUses()
    {
        NumberOfUses++;
    }

    public void SetDateTime()
    {
        LastUse = DateTime.Now;
    }

    public override string ToString()
    {
        return LastUse + ": uses " + NumberOfUses + ":" +
            SetupString;
    }
}
}
```

The key here is the attribute label above the class definition:

```
[Serializable]
    public class SetupData
```

The attribute is in brackets—**[Serializable]**. That's all you need to do to enable the feature. You define some public and private fields and some methods to access them. You can set a timestamp field, save a string description, and add a field that counts the number of usages. A byte array is also defined. (This is just a basic demonstration class for a serializable class to show the principles—for further details on serialization refer to the MSDN documentation.)

Implement the SerializeTest Class

Our test application serializes and deserializes the **SetupData** and reads and writes the data to a file on the SD card via the **SerializeTest** class.

This uses a default directory and filename, defined by constants. We will use the directory Setup, and the binary file will be called SetData.Bin. It can read the setup binary file only after a SD card is inserted, so it has a method **OnCardInserted**, which gets the **StorageDevice** instance passed in its function parameters. This method will check for a setup file in the correct directory. If it finds one, it reads the binary data and then uses this data to create a new **SetupData** instance, initialized with the saved values. If the file does not exist, the method creates a default instance of the setup data; it will also ensure that the required directory is on the SD card.

Once the new setup data instance is created, it is available for use by the application. Another method is used to handle the card being removed; this will prevent you from trying to read a card that does not exist. In our example, the main Program.cs file will include the handler for the card inserted events and will inform the test application class when a card has been inserted or ejected.

The other methods in our test application give us access to the current setup data instance and allow us to update the data and also to save the new data back to the SD card. We can also delete the binary file on the SD file, giving us a clean setup next time the board is reset.

The following is the code for the **SerializeTest** class:

```
using System.IO;
using Gadgeteer;
using Microsoft.SPOT;

namespace SerializeSD
{
    public class SerializeTest
    {
        private string m_filePath;
        private SetupData m_setUpData;
        private StorageDevice m_sd;
        public const string DEFAULT_DIR = "Setup";
        public const string DEFAULT_FILE = "SetData.bin";
        private bool m_cardInserted = false;

        public SerializeTest()
        {
            m_filePath = DEFAULT_DIR + "\\" + DEFAULT_FILE;
            // create a default instance of data
            m_setUpData = new SetupData("first use");
        }

        public void OnCardInserted(StorageDevice sd)
        {
            m_sd = sd;
            if (File.Exists(m_sd.RootDirectory + "\\" + m_filePath))
            {// we have a save setup file
                byte[] buffer = m_sd.ReadFile(m_filePath);
                m_setUpData = (SetupData)Reflection.Deserialize(buffer,
                                            typeof(SetupData));
            }
            else
            { // no file so create new setup data
                if (Directory.Exists(sd.RootDirectory + "\\" +
                    DEFAULT_DIR) == false)
                {// no directory for data file - so create it
                    m_sd.CreateDirectory(DEFAULT_DIR);
```

```
                }
            }
            m_cardInserted = true;
        }

        public void OnCardEjected()
        {
            m_cardInserted = false;
            m_sd = null;
        }

        /// <summary>
        /// Get the setup data
        /// </summary>
        public SetupData GetSetup
        {
            get { return m_setUpData; }
        }

        /// <summary>
        /// Save setup to SD
        /// Only if we have an SD Card
        /// </summary>
        /// <returns>true if success</returns>
        public bool SaveSetupData()
        {

            if (m_cardInserted)
            { // only if an SD card inserted
                byte[] buffer = Reflection.Serialize(m_setUpData,
                    typeof(SetupData));
                m_sd.WriteFile(m_filePath, buffer);
            }

            return m_cardInserted;
        }

        /// <summary>
        /// delete any saved setup file
        /// </summary>
        public void DeleteSetupFile()
        {
            if (m_cardInserted)
            {
                File.Delete(m_filePath);
            }
        }
    }
}
```

The two key sections read the binary file, then use the data to create a new **SetupData** instance from the data (deserialize), then serialize the data to create a byte array of all the field values (serialize), and then save this data buffer to the file. The read and create new data code is in the **OnCardInserted()** function. We then check to see if the correct file is on the SD card; if the file exists, we read the binary file using the **StorageDevice ReadFile** function. This gives us a byte array with the required data. We then use this data buffer to create a new instance of the **SetupData** class, using the .NET **Deserialize** function, as in the following code snippet:

```
byte[] buffer = m_sd.ReadFile(m_filePath);
m_setUpData = (SetupData)Reflection.Deserialize(buffer,
                                    typeof(SetupData));
```

The first line reads the binary data from the file, as discussed earlier. We use the **Reflection.Deserialize** function to create our class instance. We pass it the data to use and tell it the type of class to create. The function will return a type of object, even though we have told it the type of class to create, so we also need to cast the result to a **SetupData** class—using the casting expression (**SetupData**). Our **m_setUpData** is now a new instance of the class, populated with our saved data.

The saving of the **SetupData** class to a file is in the function **SaveSetupData**. This will use the **Reflection.Serialize** function to create a byte array of our data fields, which we can then save as a binary file to the SD card, as in the following code:

```
public bool SaveSetupData()
    {
        if (m_cardInserted)
        { // only if an SD card inserted
        byte[] buffer = Reflection.Serialize(m_setUpData,
        typeof(SetupData));
            m_sd.WriteFile(m_filePath, buffer);
        }
        return m_cardInserted;
    }
```

Once again, when we call the serialize function, we need to tell it the type of class we are converting and give it the actual instance of the class we want to convert. It is then a simple matter of saving the byte array to the SD card using the **StorageDevice WriteFile** function.

The Program.cs File

The final part is the Program.cs file. We need to add the event handler for the card insertion/removal event and from the card insert event inform our test application, which will then load the setup data from the file. After this, we do a test of the data, printing its key values to the debug text output; then we update the setup values and save them back to the file.

In this example, every time the SD card is inserted, the new setup data is used and modified. As mentioned earlier, if the SD card is already inserted on power-up, you most likely will not get a card inserted event. To handle this instance, we check whether a card is inserted when the application starts. If a card is already inserted, we use a timer to delay handling this until after the application framework has been started; when the timer fires, we treat it like a card has just been inserted.

The event handler for card inserted and card ejected will call the relevant functions on our **SerializeTest** class.

The code for Program.cs is as follows:

```
using Microsoft.SPOT;
using GT = Gadgeteer;
using Gadgeteer.Modules.Sytech;
using Timer = Gadgeteer.Timer;

namespace SerializeSD
{
    public partial class Program
    {
        private SerializeTest m_testApp;

        // This method is run when the mainboard is powered up or reset.
        void ProgramStarted()
        {
            m_testApp = new SerializeTest();
            ethernetSD.OnMediaChanged += new EthernetSD.
                MediaChangeHandler(ethernetSD_OnMediaChanged);
            if (ethernetSD.CardInserted)
            { // card inserted on power up
                m_testApp.OnCardInserted(ethernetSD.SDCard);
                // get our set up data after app is running as card
                // inserted
                GT.Timer timer = new GT.Timer(500, Timer.BehaviorType.
                    RunOnce); // every second (1000ms)
                timer.Tick += new Timer.TickEventHandler(timer_Tick);
                timer.Start();
            }

            // Use Debug.Print to show messages in Visual Studio's
            // "Output" window during debugging.
            Debug.Print("Program Started");
        }

        void timer_Tick(Timer timer)
        {
            TestConfiguration();
        }

        private void TestConfiguration()
        {
            SetupData data = m_testApp.GetSetup;
            Debug.Print(data.ToString());
            UpdateConfiguration();
        }

        private void UpdateConfiguration()
        {
            SetupData data = m_testApp.GetSetup;

            //put some test data in our setup array
            byte[] testarray = new byte[10];
            for (int x = 0; x < 10; x++)
            {
```

```
            testarray[x] = (byte)(x + 10);
        }
        data.TestArray = testarray;
        data.IncUses();
        int testint = data.SetupInt;

        data.SetupString = "change the string:" + testint++;
        data.SetupInt = testint;
        data.SetDateTime();
        //now save it
        m_testApp.SaveSetupData();
        Debug.Print("Setup Data Updated");
    }

    void ethernetSD_OnMediaChanged(object sender,
                                   GT.StorageDevice sdCard,
                                   bool cardInserted)
    {
        if (cardInserted)
        {
            Debug.Print("card inserted");
            m_testApp.OnCardInserted(sdCard);
            TestConfiguration();
        }
        else
        {
            Debug.Print("card removed");
            m_testApp.OnCardEjected();
        }
    }


    }
}
```

NOTE *This code is for a Sytech NANO mainboard. If you are using a GHI-based mainboard the code is very similar, but separate events are used for card inserted and card ejected, as discussed at the start of the chapter.*

If you build and run this on your hardware, it works as follows.

When an SD card is detected, it is checked to determine whether it has a binary setup file. If not, a default setup data instance is created. If it has a binary file, this is used to load the setup data. It will then print out in the debug text window the key values of the setup data. We then update the setup data, putting in a new timestamp value; increment the number of uses value; and add some dummy data to the byte array. The new setup data is then saved to the SD card.

If the SD card is removed and then replaced, the process is repeated, showing how the file has been updated. If the device is powered off or reset, then the last setup data saved to file will be used to initialize the setup data when the SD card is detected.

This project has demonstrated how you can persist application data to a file and then re-create the data object from the file. This allows quite complex data structures to be saved and restored in a fairly simple manner. Being able to save and restore data is a fairly fundamental requirement for embedded projects.

Micro Framework Extended Weak References

An alternative way of saving and restoring application data without an SD card is to use *.NET Micro Framework Extended Weak References*. This is also a serialization-based method that saves the data object to the mainboard's Flash memory. However, the mainboard needs to be able to supply an area of Flash memory for this usage. Not all mainboards do this (those with limited amounts of Flash memory may not). Also, different mainboards will provide different amounts of Flash for this purpose—in general, the area is quite small, and if you try to save more data than the area can hold, it will overwrite some of the old data. Using the SD card with serialization offers a more mainboard independent and flexible solution to the problem. It also allows you to take data created by one device, save it to SD, remove the SD card, and insert it in a different device (running compatible software) and use the same data.

Text and CSV File Projects

Now let's look at using text-based files and using standard format text formats for reading and writing records. Text files contain lines of readable characters. Each line is terminated with carriage return and line feed characters. When we read and write text files, we deal with strings rather than arrays of bytes. In particular, we will look at a formatted text file used for records of data: a CSV (comma-separated value) file.

A CSV file is a text-based format for storing tabular data. Most spreadsheet applications support importing and exporting data in a CSV format. Each line is a record (or spreadsheet row). Each data value (or column) in the row is represented in plain text form, and each value is separated by a comma character. For instance, a CSV file may contain a date, amount, and quantity, like this:

```
"24/07/12,47.99,100<cr><lf>"
```

Simple Text Logger Project

In this project, we will create a simple text logger that will allow you to open a text file and add logging strings. Each logging string will have a timestamp prepended to each entry.

If the file exists, the new entries will be added to the end. In a real logger, you would monitor the size of the file, and when it exceeds a defined size, the file would close and a new one would open, preventing a logging file from getting unmanageably large. In general, a file is not written until the file is closed, even if you flush each entry after write, so you do need to close the file to preserve the data. In our example, we will add ten log entries and then close the file. A more secure method is to open the file in append mode, add the entry, and then close the file. Opening and closing the file for each entry will take longer, but it ensures a more secure, robust file write.

We use a .NET **StreamWriter** class to write our text file and open a text stream associated with a physical file. You can open the file in create, overwrite, or append mode. In append, the existing file is opened and any new text is added to the end of the file. We use the **StreamWriter.WriteLine** function to add our string to the file. This function will add a line terminator to the end of our text. The default is carriage return/ line feed. You can change the terminating character(s) using the **StreamWriter.NewLine** property, passing a string with the terminating characters included.

Add a new Gadgeteer project called TextFile. Add a mainboard and SD module. For our example, we are using a Sytech NANO and SD module. Add a new class to the project called **Logger.cs**. This will be our text logger class.

Logger.cs

Add the following code to the Logger.cs file:

```
using System;
using System.IO;

namespace TextFiles
{
    /// <summary>
    /// Simple text logger class
    /// </summary>
    public class Logger
    {
        private StreamWriter m_logFile;
        public bool FileOpen { get; private set; }
        public Logger(string path, string logname)
        {
            if (Directory.Exists(path) == false)
            { // create the directory
                Directory.CreateDirectory(path);
            }
            string filename = path + "\\" + logname;
            m_logFile = new StreamWriter(filename,true);
            FileOpen = true;
            AddEntry("logger opened");
        }

        public void AddEntry(string logEntry)
        {
            if (FileOpen)
            {
                string entry = DateTime.Now.ToString() + ":" +
                               logEntry;
                m_logFile.WriteLine(entry);
                m_logFile.Flush();
            }
        }

        public void CloseLog()
        {
            if (FileOpen)
            {
                AddEntry("logger closed");
                FileOpen = false;
                m_logFile.Close();
                m_logFile.Dispose();
            }
```

```
            }
        }
    }
```

In the constructor, we pass in the directory path (the full path including the root directory) and the filename we want to use for the logger. We check whether the directory exists; if not, we create it. We then create a new StreamWriter instance, using the path and filename of our logger. We then add an initial entry to the logger to mark the file opening.

Our **AddEntry** function takes the string we want to add to the logger as a parameter. We get the current data and time to create a timestamp and prepend it to our string. We then call the **StreamWriter.WriteLine** function to add the string to the file. The line terminating characters will be added to the end of our string (the default is **<cr><lf>**). We call the flush function to try and write the entry to the file now; however, most hardware currently will not write the file completely until the file is closed.

Our final function is the **CloseLog** function. This will close the StreamWriter after adding a file closed log entry.

Program.cs

Our test application is in the Program.cs file. We will add event handlers to detect a card inserted. When a card is inserted, we will create our simple logger class and then use the timer to write an entry to the logger every 0.5 seconds. After ten entries have been written, we will close the file. If we remove the card and then insert it again, we will open the log file and repeat the process, but this time the log entries will be appended to the end of the previous session.

Add the following code to the Program.cs file:

```
using Gadgeteer;
using Microsoft.SPOT;
using GT = Gadgeteer;
using Gadgeteer.Modules.Sytech;
using Timer = Gadgeteer.Timer;

namespace TextFiles
{
    public partial class Program
    {
        private Logger simpleLogger;
        private int entries;
        private GT.Timer timer;

        // This method is run when the mainboard is powered up or reset.
        void ProgramStarted()
        {
            ethernetSD.OnMediaChanged += new EthernetSD.
                MediaChangeHandler(ethernetSD_OnMediaChanged);
            timer = new Timer(500);
            timer.Tick += new Timer.TickEventHandler(timer_Tick);
            entries = 0;
            if ( ethernetSD.CardInserted)
            {
                InitLogger(ethernetSD.SDCard);
```

```
        }

        // Use Debug.Print to show messages in Visual Studio's
        // "Output" window during debugging.
        Debug.Print("Program Started");
    }

    void timer_Tick(Timer timer)
    {
        entries++;
        if (entries < 10)
        {
            simpleLogger.AddEntry("Log entry :" + entries);
        }
        else
        {
            timer.Stop();
            entries = 0;
            simpleLogger.CloseLog();
            Debug.Print("close log file");
        }
    }

    void InitLogger(StorageDevice sd )
    {
        string dir = sd.RootDirectory + "\\logDir";

        simpleLogger = new Logger(dir, "logger01.txt");
        timer.Start();
        Debug.Print("Open log file");
    }

    void ethernetSD_OnMediaChanged(object sender,
                            GT.StorageDevice sdCard, bool
                                cardInserted)
    {
        if (cardInserted)
        {
            InitLogger(sdCard);
        }
    }
    }
}
}
```

The code is very similar to that of our previous projects. We attach event handlers to the card detect event. We also have to check whether a card is already inserted when the application starts, because we will not see a card insert event. If a card is inserted or a card is already inserted on startup, we call the **InitLogger** function. This will define the logging directory to use and create a new logger, passing in the directory and logger filename to use. It will then start the timer, which will trigger every 0.5 seconds. The timer handler will generate a logging entry and a number and then write this string to the logger. After ten entries have been written (about 5 seconds), the logger is closed.

If you build and run this application, after you insert the SD card, ten logging entries will be written. If you remove and reinsert the card, ten more will be added. If you remove the card and use a card reader to read the card on your PC, in the logDir, you will find a file called logger01.txt. If you open this file (it's just a text file), you will see the logging entries. The following is an example of the contents of a logging file:

```
01/01/2009 00:00:29:logger opened
01/01/2009 00:00:30:Log entry :1
01/01/2009 00:00:30:Log entry :2
01/01/2009 00:00:31:Log entry :3
01/01/2009 00:00:31:Log entry :4
01/01/2009 00:00:32:Log entry :5
01/01/2009 00:00:32:Log entry :6
01/01/2009 00:00:33:Log entry :7
01/01/2009 00:00:33:Log entry :8
01/01/2009 00:00:34:Log entry :9
01/01/2009 00:00:34:logger closed
```

CSV File Project

In this project we will use CSV files to load data from the SD card into the application. The application can also save data records to the file. This allows data records to be simply loaded from a file on the SD card by exporting the data from a spreadsheet or other desktop application.

We will load customer loyalty card records into an array of records. Each record has the card number, the points limit for the card, the current points, and the customer name. When a customer presents his or her loyalty card to the application (pretend we have a serial card reader connected to our hardware), the card number can be read from the card and then used to look up the customer details from our loaded records.

Another example of CSV file usage is in a GPS logger system. The system reads the GPS location every minute, and it also reads some other data such as ambient temperature, atmospheric pressure, and so on. It then saves all this as a CSV data record to the SD card. At the end of the day, the SD card is removed and the data can then be read into a spreadsheet application and reports can be generated.

This time, when writing a data record, we open the file, write the record, and then close the file.

The data records are formatted as strings, with a comma separating each field. To read and write file records, we use a StreamWriter, as the data is text-based. When we read a record, each field is separated by a comma character. We can use the string function **Split** to separate each field from the string, using the comma as a marker. This gives us an array of strings, with an element for each field. Note that we can use other characters or character combinations to define the separator.

To encode our record, we generate a string by adding each field value separated by a comma.

Keeping to our object-oriented principles, we create a class for the record. This class is responsible for encoding and decoding the fields to/from a CSV string and making each field available. If the format of our records changes, we just have to modify this one class.

Create a new Gadgeteer project called CSVApp. Add a mainboard and SD module; we will use a Sytech NANO and SD module in this example.

Add a Record.cs Class

Add a new class to the project called Record.cs. This will encapsulate our example loyalty card record. The record will have a string card number, an integer value for the maximum allowed points, an integer current points value, and a string with the customer name. The record will be encoded as *CardNumber,Max_Value,Points,CustomerName*. All fields are represented as string values, so the two integers will be supplied as the string values of the integers. Our constructor will allow a default record, with zero values and a record to be constructed from a CSV-encoded string. We also have a function to allow the record values to be loaded or modified from a CSV string. Note that we have to convert the string representations of the integer values to actual integers in this process.

We can read the actual field values from properties and also get a CSV-encoded string of the values. Finally, for logging purposes, we have a **ToString** function showing the contents of each field. The code for the class is fairly simple and self-explanatory. The following code makes up our record class:

```
namespace CSVApp
{
    public class Record
    {
        public string CardNumber { private set; get; }
        public uint Max_value { private set; get; }
        public uint Points { private set; get; }
        public string Cust_name { private set; get; }

        public Record()
        {
            CardNumber = "UnKnown";
            Max_value = 0;
            Points = 0;
            Cust_name = "Unknown";
        }
        public Record(string csvEntry):this()
        {
            LoadfromCSV(csvEntry);
        }

        public void LoadfromCSV(string csvEntry)
        {
            string[] fields = csvEntry.Split(new char[] {','});
            // simple error check
            if (fields.Length == 4)
            {
                CardNumber = fields[0];
                Max_value = uint.Parse(fields[1]);
                Points = uint.Parse(fields[2]);
                Cust_name = fields[3];
            }
        }

        public string ToCSV()
        {
```

```
            string csv = CardNumber + "," +
                            Max_value.ToString() + "," +
                            Points.ToString() + "," +
                            Cust_name;
            return csv;
        }

        public override string ToString()
        {
            return CardNumber + ":" +
                            Max_value.ToString() + ":" +
                            Points.ToString() + ":" +
                            Cust_name;
        }
    }
}
```

Add a CSVHandler.cs Class

Add a class to the project called **CSVHandler.cs**. This class will be responsible for reading and writing records to our CSV file. It will also maintain the **ArrayList** of records loaded from the CSV file.

An array list is a collection of objects. Unlike normal fixed arrays, the size of the array list (number of items) is dynamic, so it grows as you add more items. It provides functions to read the records from a CSV file and load them into our array list. The class provides the array list of records, as a property.

The function **LoadFromFile()** will load the array list of records from the CSV file, passed in the path parameter. It will open the file as a StreamReader, read each line (record) from the file, and use the CSV string to create a new record instance. The record instance is then added to the array list. It returns the number of records read from the file.

The function **AddRecordToFile()** will add the record passed to the file, passed as a path. It will open the file using a StreamWriter (in append mode) and get the CSV string from the record instance and write this to the file. The file is closed after writing.

The code for the **CSVHandler.cs** class is as follows:

```
using System;
using System.Collections;
using System.IO;
using Microsoft.SPOT;

namespace CSVApp
{
    public class CSVHandler
    {
        public ArrayList Records { private set; get; }

        public CSVHandler()
        {
            Records = new ArrayList();
        }

        public int LoadFromFile(string filepath)
        {
```

```csharp
        int items = 0;

        if (File.Exists(filepath))
        {
            string csvrecord;
            StreamReader csvFile = new StreamReader(filepath);
            bool hasdata = true;
            using(csvFile)
            {
                do
                {
                    csvrecord = csvFile.ReadLine();
                    Record nxRecord = new Record(csvrecord);
                    Records.Add(nxRecord);
                } while (!csvFile.EndOfStream);
            }
            items = Records.Count;
        }
        return items;
    }

    public bool AddRecordToFile(Record record, string filePath)
    {
        bool success = false;
        try
        {
            StreamWriter csvFile = new StreamWriter(filePath,true);
            csvFile.WriteLine(record.ToCSV());
            csvFile.Close();
            csvFile.Dispose();
            success = true;
        }
        catch (Exception)
        {
            Debug.Print("Error writing to CSV File");
        }
        return success;
    }
  }
}
```

Program.cs

The final part of the application is in the Program.cs file. We will add a handler to the card inserted event of the SD module to detect when a card is inserted. As in previous projects, we will also check a card inserted on application startup. When a card is detected, we run our test code.

The code for the Program.cs file is as follows:

```csharp
using System.Collections;
using System.IO;
using Gadgeteer;
using Microsoft.SPOT;
using GT = Gadgeteer;
```

```
using Gadgeteer.Modules.Sytech;
using Timer = Gadgeteer.Timer;

namespace CSVApp
{
    public partial class Program
    {
        private CSVHandler fileHandler;
        private const string DIR_NAME = "\\CSVDir";
        public const string FILE_NAME = "\\CSVRecs.csv";
        // This method is run when the mainboard is powered up or reset.
        void ProgramStarted()
        {
            fileHandler = new CSVHandler();
            ethernetSD.OnMediaChanged += new EthernetSD.
                MediaChangeHandler(ethernetSD_OnMediaChanged);
            if (ethernetSD.CardInserted)
            { // card already inserted
                GT.Timer timer = new Timer(500,Timer.BehaviorType.
                    RunOnce);
                timer.Tick += new Timer.TickEventHandler(timer_Tick);
                timer.Start();
            }
            // Use Debug.Print to show messages in Visual Studio's
            // "Output" window during debugging.
            Debug.Print("Program Started");
        }

        void timer_Tick(Timer timer)
        {
            TestCSVFile(ethernetSD.SDCard);
        }

        void ethernetSD_OnMediaChanged(object sender,
                                    GT.StorageDevice sdCard, bool
                                    cardInserted)
        {
            if (cardInserted)
            {
                TestCSVFile(sdCard);
            }
        }

        private void TestCSVFile(StorageDevice sdCard)
        {
            string directory = sdCard.RootDirectory + DIR_NAME;
            if (Directory.Exists(directory) == false)
            {
                Directory.CreateDirectory(directory);
            }
            string path = directory + FILE_NAME;
            if ( File.Exists(path)== false)
            {// create a dummy csv file
                CreatTestCSV(path);
            }
```

```
    else
    { // file exists  - so load it
        fileHandler = new CSVHandler();
        int numRecords = fileHandler.LoadFromFile(path);

        Debug.Print("Loaded " + numRecords + " from file");
        ArrayList list = fileHandler.Records;
        foreach (Record item in list)
        {
            Debug.Print(item.ToString());
        }
    }
}


/// <summary>
/// Example of writing CSV records
/// to a file, used to create a test
/// file none on SD card
/// </summary>
/// <param name="path"></param>
private void CreatTestCSV(string path)
{
    Debug.Print("Creating test CSV file");
    CSVHandler testhandler = new CSVHandler();
    Record testRecord = new Record("123456,10000,345,Fred J");
    testhandler.AddRecordToFile(testRecord, path);
    testRecord.LoadfromCSV("1239996,10000,105,John G");
    testhandler.AddRecordToFile(testRecord, path);
    testRecord.LoadfromCSV("1786555,5000,96,Sarah P");
    testhandler.AddRecordToFile(testRecord, path);
    Debug.Print("Test file created");

    }
  }
}
```

Our test code will check to see if a CSV file is in the CSV directory. We set the name of the directory and file with string constants. If the file does not exist, we create a test CSV file and write it to the card; this gives an example of writing CSV files. The function **CreateTestCSV(*path*)** is responsible for this. It will create three test records with data in them and then write these to a CSV file.

If when a card is inserted, the CSV file exists, then the **CSVHandler** class is used to load the records in the file into our record array. We then iterate through the record array and print the details of each record to the debug output.

The first time you run the application, the CSV file will not normally be on the SD card, unless you have created one and put it there. So the first time around, the application will create a test file. If you remove and insert the SD card after this, the new CSV file will be detected and used to load the record array.

You can remove the SD card and insert it into a PC reader. If you navigate to CSVDir\CSVRecs.csv, you can open the file (it's a text file), and you will see each comma-separated record. You can also open the file with a spreadsheet application such as Excel and see the records. Add some new records using either a text editor or your spreadsheet, and then save the file (keep it as a CSV file in Excel; don't convert it to a worksheet). Now if you remove the SD card from the PC reader (ejecting it first), you can read the modified records in our test application. Run the application either from Visual Studio, so you can see the debug output, or from MFDeploy, using the device connect function to see the debug text. Now when you insert the card, the CSV file will be read, the records extracted, and then the record summary printed in the debug window.

Summary

In this chapter we have explored the theory behind using the SD card module and how the Gadgeteer **StorageDevice** class works. We have also seen how to extend the functionality provided by the **StorageDevice** class to enable direct handling of file functions.

We have looked at how we can persist application data using binary files, allowing data to be available for the application's next session after a power-down cycle.

Finally, we looked as using text-based files instead of binary files. We looked at how we could use text files to create a simple text logger and also how we can use CSV-encoded text files to import and export data from the application to spreadsheets and database applications.

CHAPTER 11

Ethernet and Web Device Projects

I n this chapter we will look at using Ethernet-connected projects. We will examine the basic principles of communicating using TCP/IP servers and clients; User Datagram Protocol (UDP) and web devices; and web servers and web clients.

Gadgeteer mainboards that support Ethernet either have the Ethernet hardware integrated into the mainboard or use an external Ethernet hardware device on a module, usually interfaced to the mainboard using a Serial Peripheral Interface (SPI) connection. The low-level native driver interface between the processor and the .NET Micro Framework is specific to the mainboard hardware. If the Ethernet hardware is integrated into the mainboard, the Ethernet module is usually just the RJ-45 connector. In general, Ethernet hardware modules are manufacturer specific—that is, you need to use the module that was designed for your specific mainboard.

Ethernet SPI Modules
When the Ethernet hardware is incorporated into an external module, most manufacturers use the same Microchip (SPI Ethernet chip) part. This provides the Ethernet functionality in a single chip, with an SPI interface to the processor. There have been sample drivers for this part in the Micro Framework Porting Kit since Ethernet was first supported—hence the reason for the common use of this hardware. However, the pin-outs of the cables from the module to the mainboard are usually different from manufacturer to manufacturer, mainly regarding which pin is the interrupt and which is the chip select and reset. So even though the hardware is almost identical, the pin usage on the Gadgeteer socket is not. You need to check which Ethernet modules your mainboard supports before buying modules.

From an application software interface point of view, they all have the same .NET Micro Framework interface. Each manufacturer may incorporate a few additions to the interface, such as different ways to detect whether a network connection exists.

In our code examples, we will use only .NET Micro Framework API functions, so the code should work on any mainboard.

To use Ethernet in your application, you do not actually access the Ethernet hardware directly at all; instead, you use sockets. Do not confuse these with Gadgeteer sockets. A

Gadgeteer socket encapsulates a physical connection socket from the mainboard that is used to connect to modules. A socket encapsulates an endpoint of an interprocess communication flow across a computer network connection. These *Internet socket* connections are based on the Internet Protocol (IP). The .NET API functions for using sockets are in the System.NET.Sockets and System.NET libraries. Both namespaces are in System.dll, which is automatically added to your application when you use a Visual Studio Gadgeteer application template. These are based on the desktop .NET functions.

NOTE *In this chapter, when I use the term "socket," I am referring to an internet socket, unless otherwise stated.*

Sockets

The socket is the basic element of any Internet communication. A socket is a connection endpoint. Sockets allow you to send and receive data across a network. Each socket has a unique address on the network. This address is an IP address, which is a 32-bit binary number. These 32-bit numbers are usually written in a human-readable form of four numbers between 0 and 255, separated by a dot—for example, 192.168.1.100 (a little simpler than the hexadecimal form 0xc0a80164!).

To send and receive data to and from another socket (for instance, the Ethernet connection on your PC), you need to know the destination endpoint address (IP address). On the protocol side, this address can have a number of channels, or *ports*. So a complete address comprises the IP address and a port number. Some port numbers are standardized, such as HTTP port 80 (used for Internet browsing), FTP (File Transfer Protocol) port 21, and Telnet port 23.

A client application must know the IP address of the server and then connect with the server on a particular port. The server will listen for connection requests on the defined port and accept connections from clients requesting a connection on this port. Note that a server can listen for connections on a number of ports and accept connections from multiple clients.

To use our mainboard on a network, we first need to configure the network settings for our device's Ethernet adapter. These settings are stored in the nonvolatile Flash memory of the device.

Sockets allow transfer of data between endpoints on the network. So we need an endpoint for our device, and this is the IP address of our device. Remember that we don't directly access the Ethernet hardware from our application; this is how the magic happens. We assign a physical IP address to our hardware. When we create a socket class instance (using the Micro Framework API functions), we associate, or bind, the socket class instance to an endpoint by defining the IP address and port to use. The low-level drivers know which physical adapters we have and their IP addresses, so when we set the endpoint IP address in the socket class, the low-level code knows from this which physical adapter to bind the socket to. Our next step is to configure the network settings on our Ethernet module.

NOTE *In general, only one Ethernet type adapter is on our Micro Framework board. As a shortcut, we don't even specify the local endpoint when we bind a socket but let the system default to the first adapter it finds—which will be our Ethernet adapter.*

Device Network Configuration

Let's assume the device will be connected to the same network your desktop PC is on. We need to determine these settings.

1. On your PC, open the command prompt by choosing Start | All Programs | Accessories | Command Prompt:

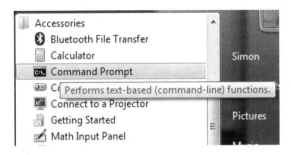

Alternatively, if you see a Run option in your Start menu, select this and enter **cmd** at the prompt.

2. In the Command Prompt window type **ipconfig** and press RETURN. This will display the IP connection information for all your network related devices, as shown next. We are interested in your Ethernet adapter.

```
C:\Windows\system32\cmd.exe

   Connection-specific DNS Suffix  . :

Ethernet adapter Local Area Connection:

   Connection-specific DNS Suffix  . : home
   Link-local IPv6 Address . . . . . : fe80::d13e:1c75:b2a0:f4b8%10
   IPv4 Address. . . . . . . . . . . : 192.168.1.105
   Subnet Mask . . . . . . . . . . . : 255.255.255.0
   Default Gateway . . . . . . . . . : 192.168.1.254
```

The IPv4 Address is the IP address of your PC local network connection. You need to know this to connect to your PC. The Subnet Mask is a filter to help simplify the number of possible addresses; the normal setting is 255.255.255.0. This means that the first three digits of an IP address must match and only the fourth digit is used for variable addresses. So in our example, the IP address is 192.168.1.105, and we will consider only addresses that are 192.168.1.*xxx*. The default gateway is the IP address of the device controlling the network; this is normally your router (or in most cases your broadband connection/router).

We need to set our device's IP address to be compatible with your local network— that is, it needs to be 192.168.1.*xxx* in our example. The *xxx* value needs to be an unused address on your network because network device addresses need to be unique. Normally, the addresses are assigned automatically by your router, but we want to assign a static address to our device so we have more control over it. Suppose that *140* is an unused address (a reasonable guess as our PC is 105 and it is unlikely we have another 35 devices on the network). To test whether it is free, type the **ping 192.168.1.140** command into the command prompt window and press RETURN.

This will launch a network utility that basically asks for anything on this address to please respond. If no device is using this address, the results from the ping will look something like this:

```
C:\Users\Simon>ping 192.168.1.140
Pinging 192.168.1.140 with 32 bytes of data:
Request timed out.
Request timed out.
Request timed out.
Request timed out.
Ping statistics for 192.168.1.140:
    Packets: Sent = 4, Received = 0, Lost = 4 (100% loss),
```

This shows that no device responded, meaning we can use this address.

We can use MFDeploy to configure our device's network settings. Connect your device via USB to your PC and start up MFDeploy. Set MFDeploy to detect USB devices, and you should see your device's ID in the Device Edit box. From the menu tab, select Target | Configuration | Network, as shown next:

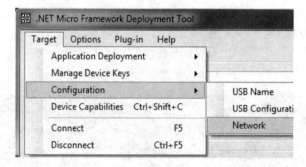

After a few seconds, the Network Configuration dialog will open, showing the current network settings. Figure 11-1 shows the dialog with the settings after we make the changes discussed next.

FIGURE 11-1   Network Configuration settings

Our tests showed we can use 192.168.1.140 as our network address, or Static IP Address. We will use a Subnet Mask of 255.255.255.0 and the same Default Gateway as our PC—192.168.1.254. Enter these into the relevant boxes in the Network Configuration dialog. The DNS setting is the address of a name server, which provides lookup capability for URL names, such as www.google.com, and will return the IP address—making it easy for a human's touch. We will use our gateway server as the DNS Primary Address. The MAC address is a unique number to identify our Ethernet hardware. Every adapter in the world should have a unique address. The addresses are controlled by a central registry, and manufacturers buy a block of addresses for their use. Your Ethernet hardware should have had a unique MAC address supplied by the manufacturer. It is most likely on a label on the board or supplied separately. The DHCP check box allows you to enable automatic address setting by your DHCP server (usually your broadband router). This allows a central device to control the network address settings and assign addresses to devices. Leave this unchecked, because we want to control the IP address used.

Below this is the section for setting WLAN (Wireless LAN) settings. We are not concerned with this section because we are not using WLAN. When you have finished the Network section of the dialog, it will look like Figure 11-1. Note that these settings are for my network; you may need different addresses for your own, depending on what you found when you used **ipconfig**.

The final step is to click the Update button. This will write the new settings to your device's Flash memory. Some devices might need their reset buttons pressed to restart the OS after this operation, but most don't.

NOTE *The network settings are actually stored on the mainboard for use by the Micro Framework OS. The low-level drivers will configure the Ethernet hardware during boot-up. This means the MAC address and so on are stored on the mainboard. If your Ethernet hardware is on a separate module (that is, one of the SPI-based setups), the MAC address is associated with the mainboard and not the adapter. So if you plug a different Ethernet module into the hardware, it will use the same settings.*

Now that we have the network settings configured for our device, we need to test them.

1. Connect a network cable to your device's Ethernet (you did not need a network connection to set up the settings). If your Ethernet socket (RJ-45) has some LEDs, you should see them flash occasionally.

2. Go back to the command window on your PC, or open a new one if it has been closed.

3. Repeat the previous test, using **ping**, to check out our new IP address. Type **ping 192.168.1.140** and press RETURN. If your device has a different IP address, use that one. You should see something like this:

```
C:\Users\Simon>ping 192.168.1.140
Pinging 192.168.1.140 with 32 bytes of data:
Reply from 192.168.1.140: bytes=32 time<1ms TTL=255
Reply from 192.168.1.140: bytes=32 time<1ms TTL=255
Reply from 192.168.1.140: bytes=32 time<1ms TTL=255
```

```
Reply from 192.168.1.140: bytes=32 time<1ms TTL=255

Ping statistics for 192.168.1.140:
    Packets: Sent = 4, Received = 4, Lost = 0 (0% loss),
Approximate round trip times in milli-seconds:
    Minimum = 0ms, Maximum = 0ms, Average = 0ms
```

This means the device is configured correctly for our network and is responding to messages. The **ping** functionality is built into the mainboard OS network drivers, so you do not need any application running for it to work—just the TinyCLR (main OS) will be running.

TCP/IP Server Project

For our network example projects, we'll use the Mountaineer Ethernet mainboard. This mainboard is slightly different from other Gadgeteer mainboards in that it includes the USB module and Ethernet module integrated onto the mainboard. The Ethernet module is still added using the GUI designer; the difference is that there is no physical Gadgeteer socket for the Ethernet module, so you just drag-and-drop the Ethernet module from the Toolbox, but you do not need to connect it because this is done automatically. The framework still returns an Ethernet module class, so from an application point of view, there is no difference. The actual example code will work on any mainboard and Ethernet module (or even a pure Micro Framework board).

TCP/IP is a software-level protocol, and the socket class (.NET IP socket) handles all the low-level protocol implementation using the OS TCP/IP stack. When the connection between the two endpoints is established, you can read and write a data stream (an array of bytes) sent and received between the two endpoints.

TCP/IP is a connected protocol. One of the endpoints is a TCP/IP server, and the other is a client. The server will listen for connection requests from clients, using its IP address and a specified port number. The client must have access to the server network and know the server's IP address and the port number in use. The client will request a connection from the server. When the server receives and accepts the connection request, it will create a new socket for the session with the client. The client will then communicate with this new session socket. The original server socket will then return to listening for connection requests from other clients.

While a socket is listening for network messages, the call to listen will block the thread. So if you call the listen function on the main thread, you will block the thread and prevent the OS and the Gadgeteer framework from operating correctly. In our server example, we will create a thread for the server to listen on, freeing up the main thread. When a connection request is received, the session socket will have a new thread created and operate from there.

To test our server, we need a client application running to request connections and transfer our test data. We will write a simple client application to run on the PC. As the Micro Framework is a subset of the full .NET Framework, we will write our code for the PC so that it can also be used on our mainboard. All we need to change are the project references for the .NET libraries, so we use the desktop libraries rather than the Micro Framework libraries.

Start by creating a new Gadgeteer application using the Visual Studio template.

TIP *The Mountaineer board is a new board and supports only Micro Framework 4.2. To use it, you need the Micro Framework 4.2 SDK installed and the latest Gadgeteer SDK (at the time of writing, this is version 600). With MF 4.2 and Gadgeteer x.600 installed, you can have projects that support MF 4.1 or MF 4.2, to match the firmware version installed in your mainboard. The code in these projects will work on MF 4.1 or MF 4.2.*

We are using the Mountaineer mainboard, so we add this mainboard and the Mountaineer Ethernet module. If you are using a different mainboard, add this and the matching Ethernet module.

Our TCP/IP server class runs like a service; it has its own thread and, once started, sits in the background waiting for connection requests. When it receives a connection request, it will start a session with the client. This session will run in its own thread as well. Our message data will just be simple strings. When a string is received over the network from the client, we will raise an event to inform the main application of the arrival of a message. The application can send a response to the received message by setting the response message in the event. This will be sent to the client by the session thread.

Now add a new class to the project called **Server.cs**:

```
using System;
using System.Net;
using System.Net.Sockets;
using System.Threading;
using Microsoft.SPOT;

namespace TCPServer
{
    /// <summary>
    /// Simple TCP/IP Server demonstration
    /// </summary>
    public class Server
    {
        private const int SERVER_PORT = 1000;
        private Socket m_serverSkt;
        private int m_port;
        private Thread m_serverThread;
        private bool m_srvRunning;
        // On Message Received event
        public event ClientMsgRxDelegate OnMessageRx;

        public Server():this(SERVER_PORT)
        {}
        public Server(int port)
        {
            m_port = port;
            m_srvRunning = false;
            m_serverSkt = new Socket(AddressFamily.InterNetwork,
                SocketType.Stream,ProtocolType.Tcp);
        }
```

```csharp
public void Start()
{
    if (m_srvRunning)
    { // only allow one server thread at a time
        Debug.Print("Server running - only one instance
                    allowed");
        return;
    }
    m_srvRunning = true;
    //Create local endpoint and bind to socket
    IPEndPoint svrEndPoint = new IPEndPoint(IPAddress.
        Any,m_port);
    m_serverSkt.Bind(svrEndPoint);
    m_serverSkt.Listen(4);  // limit number of connections to 4
    Debug.Print("Server Listening..");
    //create and start our main server thread
    m_serverThread = new Thread(ProcessServer);
    m_serverThread.Start();
}

/// <summary>
/// Fire our message received event
/// </summary>
/// <param name="args"></param>
private void MessageRx(ClientEventArgs args)
{
    if (OnMessageRx != null)
    {
        OnMessageRx(this, args);
    }
}

/// <summary>
/// This is our main server listening thread
/// </summary>
private void ProcessServer()
{
    while (m_srvRunning)
    {
        // wait for a connection - will block
        Socket client = m_serverSkt.Accept();
        Debug.Print("Connection to client");
        // create a new session for the connection
        Clienthandler handler = new Clienthandler(this,client);
    }
}

#region client handler class
/// <summary>
/// Client socket handler helper class
/// </summary>
private class Clienthandler
{
    private Socket m_clientSkt;
    private Server m_serverSkt;
```

```csharp
private Thread m_clientThread;

private const int MICROSECS_SEC = 1000000;
private int m_readTimeout = 0;

public Clienthandler(Server server, Socket client)
{
    m_serverSkt = server;
    m_clientSkt = client;
    m_readTimeout = 5 * MICROSECS_SEC;

    ServiceClient();
}

public void ServiceClient()
{
    // create and start our session thread
    m_clientThread = new Thread(ProcessClient);
    m_clientThread.Start();

}

/// <summary>
/// The session thread
/// Listen for messages
/// </summary>
private void ProcessClient()
{
    using (m_clientSkt)
    {
        while (true)
        {
            try
            {
                if (m_clientSkt.Poll(m_readTimeout,
                    SelectMode.SelectRead))
                {
                    // if read buffer is 0, then no data
                    // after timeout
                    // socket may be gone
                    int bytesavail = m_clientSkt.Available;
                    if (bytesavail == 0)
                    { // exit thread, close socket
                        Debug.Print("Client lost, close
                                    connection");
                        break;
                    }
                    // we have a message
                    byte[] buffer = new byte[bytesavail];
                    int bytesRead = m_clientSkt.
                        Receive(buffer,
                        bytesavail, SocketFlags. None);

                    ClientEventArgs args =
```

```
                                new ClientEventArgs(m_clientSkt.
                                            LocalEndPoint,
                                            buffer);
                        // let server notify of message
                        m_serverSkt.MessageRx(args);
                        // check if a response is returned in
                        // args
                        if (args.Response != null)
                        {   // send back the response
                            m_clientSkt.Send(args.Response);
                        }
                    }
                }
                catch (Exception)
                {
                    Debug.Print("Client connection error close
                        client");
                    break;
                }
            }
        }
    }
}
    #endregion
    }
}
```

The main server code is fairly simple. We have two constructors. We need to set the port the server is going to listen on. We do not need to explicitly set the IP address, because we have only one adapter (this is the normal case in a Gadgeteer board), so the socket will automatically bind to this. The base constructor will use the default port, in our example port 1000, and the overloaded constructor allows you to pass in a port number. The constructor will create the socket used as the server socket.

When we want to start the server running, we use the **Start()** function. We bind our socket to the server endpoint, and we also limit our server to a maximum of four simultaneous connections and start the socket listening for connections. The maximum number of connections supported is dependent on the TCP/IP stack implementation used by your hardware. We create our server thread in the **Start()** function. The server thread method is **ProcessServer**. It calls the socket **Accept()** function. This will block or wait until a client requests a connection.

When a client requests a connection, a new socket (with a different port number) will be created for the session, and the function returns with the session socket. Because this method blocks, we need to run the accept function in its own thread. When we have a client connection session, we create a session with the **ClientHandler** class, passing in the client socket and also a reference to the server. The client session will now be handled by the **ClientHandler** and the server thread gets back to listening for new connection requests. The only other function the server supplies is an event to inform the main application that a message has been received and provide a mechanism for the main application to send a response to the message, back to the client. This is the **OnMessageRx** event. It is fired by the **ClientHandler** when it receives a message. The event requires a delegate and event arguments to be defined. We will add this class in next, but first let's look at the **ClientHandler** class.

An instance of the **ClientHandler** is created for each new connection received by the server. We pass in the new client session socket and a reference to the main server. We need the reference to the main server so we can fire the **OnMessageRx** event. The **ClientHandler** creates a new thread to manage receiving and sending messages to the client. This thread runs the function **ProcessClient()**.

ProcessClient() will continuously poll the socket for any data received; when data is received, it is read into a buffer. Event arguments for an **OnMessageRx** event are created and populated with the received message buffer and the event is called. Any main application code subscribing to this event can access the received message raw data buffer and also has the opportunity to return a response message. After the event is fired, the **ProcessClient()** function checks the event arguments it created, and if a response has been added it sends the response to the client.

If, as a result of polling the client socket, it is detected that the client socket connection has been lost (either closed or connection lost) or an error occurred, the client session thread is terminated. We detect connection lost, or socket closed, if we get a positive response to the poll, and the available data is 0 bytes. Errors will generate exceptions, and we catch the exceptions to detect this. Because we have wrapped the socket used in the thread code in a using statement, when the using statement finishes, the client socket will be closed and freed.

As the server class is only a message transport, we have not processed the raw message data at all, but left it as an array of bytes. In this example, the message data is a string, but we leave the conversion of the data to another application class. This way, we are not limiting the server to sending and receiving strings. The application class can convert the raw data into any format.

Our event requires a delegate to be defined, and that delegate requires a custom **EventArgs** class. We'll add a new class to the project called **ClientEventArgs**. Add the following code to define the event arguments and the delegate:

```
using System;
using System.Net;
using Microsoft.SPOT;
namespace TCPServer
{
    /// <summary>
    /// Event args for client messages received
    /// </summary>
    public class ClientEventArgs:EventArgs
    {
        public IPEndPoint ClientEndPoint { get; private set; }
        public byte[] Message { get; private set; }
        public byte[] Response { get; set; }

        public ClientEventArgs(EndPoint clientEP,byte[] message)
        {
            ClientEndPoint = (IPEndPoint)clientEP;
            if (message == null) return;
            Message = new byte[message.Length];
            message.CopyTo(Message,0);
        }
    }
}
```

```
        public delegate void ClientMsgRxDelegate(object
            sender,ClientEventArgs e);
}
```

This defines a simple **EventArgs**-derived class containing the data buffer for the message, the endpoint class with the client details, and a buffer to be used to return a response message. The delegate used by our public event is also defined here.

Now let's add the code to the main application, Program.cs, to run our server. Modify the Program.cs file with the following code:

```
using System;
using System.Text;
using Microsoft.SPOT;

namespace TCPServer
{
    public partial class Program
    {

        private Server m_tcpServer;
        private int count = 1;
        // This method is run when the mainboard is powered up or reset.
        void ProgramStarted()
        {

            string ipAddress =
                ethernetForMountaineerEth.NetworkSettings.IPAddress;
            int port = 1000;
            m_tcpServer = new Server(port);
            m_tcpServer.OnMessageRx +=
                new ClientMsgRxDelegate(m_tcpServer_OnMessageRx);
            try
            {
                m_tcpServer.Start();
                Debug.Print("Server running :" + ipAddress + ":" +
                    port);
            }
            catch (Exception)
            {

                Debug.Print("Error starting server, is ethernet
                    connected?");
            }

            // Use Debug.Print to show messages in Visual Studio's
            // "Output" window during debugging.
            Debug.Print("Program Started");
        }

        void m_tcpServer_OnMessageRx(object sender, ClientEventArgs e)
        {
            // note this is not called on main thread
            // convert byte[] to string
```

```
        string message = new string(Encoding.UTF8.GetChars
            (e.Message));
        Debug.Print("Rx Msg " + count  + ":" + message);
        Debug.Print("Client :" +  e.ClientEndPoint.Address + ":"
            + e.ClientEndPoint.Port);
        // generate a response to send back to the client
        e.Response = Encoding.UTF8.GetBytes(" Message "
            + count++ +" Received");
        }
    }
}
```

We get the device IP address from the Ethernet module; this is used to generate some debug text, telling us the IP address being used. We create an instance of our server class, setting the port to use (we are using the same port as the server default port, so we could have used the base **constructor()**). We then connect a handler to the server's **OnMessageRx** event. Finally, we start the server running. Our event handler will convert the message byte array into a string and display the string in a debug statement. We also show the client's IP address and port in a debug statement. We maintain a count of the number of messages we have received and use this number to generate a response string. We convert the string to a byte array and write it to the response value of the event arguments. This response is returned to the server, where it is sent to the client.

The code assumes that the network cable is connected. Mainboards normally supply their own property to determine whether the cable is connected, typically called something like **CableConnected**. The Gadgeteer base Ethernet module code defines two events, **NetworkUp** and **NetworkDown**, plus a couple of **NetworkConnected** properties. At the current time it is not a good idea to rely on these, however. Whether they work depends on your hardware and your network. You need to test them before deciding to use them. In general, I have found that they usually don't work reliably. This is why we wrap the **server.start** call in a try/except. If the network is not connected, we catch the exception and print a debug statement.

TCP/IP Test Client

We have a server application for our device. To test it we need a client. We will implement a simple client using a console application for a PC. We will also demonstrate how code written for the Micro Framework can also work in a desktop application, with minor changes. The full desktop version of the .NET libraries simplifies this by providing a **TCPClient** class, derived from the **Socket** class. But this class is not supported in Micro Framework, so we will use a normal socket class for our client.

Add a new project to your Visual Studio solution, but this time use the Visual C# | Windows | Console Application template, as shown in Figure 11-2.

This will add a Windows Console project to your solution, with a Program.cs file with template code in for the application. Add a class to the project called Client.cs. This will contain the code for our client socket function. The client code is very similar to the server code: we create a socket, connect it to our server's address, and then in its own thread we poll it for received messages. When we receive a message, we generate a message received event for the main application. The class also provides a send method, allowing the application to send a message to the server. The code for the client will use

FIGURE 11-2   Windows Console template

.NET functions available to the Micro Framework, and the only change we need to make between desktop and Micro Framework are the library references for the debug statements.

Add the following code to your **Client.cs** class:

```csharp
using System;
using System.Net;
using System.Net.Sockets;
using System.Text;
using System.Threading;

using System.Diagnostics;  // for desktop debug statements

namespace TCPClient
{
    public class Client
    {
        private Socket m_client;
        private IPEndPoint serverEP;
        private bool connected;

        private const int MICROSECS_SEC = 1000000;
        private int m_readTimeout = 0;

        private Thread rxThread;
        private bool running;

        public event MsgRxDelegate OnMessageRx;

        public Client(string ServerAddress, int port)
        {
            IPAddress ipAddress = IPAddress.Parse(ServerAddress);
            serverEP = new IPEndPoint(ipAddress, port);
            m_client = new Socket(AddressFamily.InterNetwork,
                SocketType.Stream,
                ProtocolType.Tcp);

            connected = false;

            m_readTimeout = 5 * MICROSECS_SEC;
```

```
    }

    public void Connect()
    {
        if (!connected)
        {
            m_client.Connect(serverEP);
            rxThread = new Thread(new ThreadStart(ProcessRx));
            running = false;
        }
    }

    /// <summary>
    /// This is our client socket 'read' thread
    /// </summary>
    private void ProcessRx()
    {
        running = true;
        while (true)
        {
          if (m_client.Poll(m_readTimeout, SelectMode.SelectRead))
              try
              {
                  {
                      int bytesavail = m_client.Available;

                      if (bytesavail == 0)
                      { // exit thread, close socket
                          Debug.Print("Server lost, close
                              connection");
                          break;
                      }
                      // we have a message
                      byte[] buffer = new byte[bytesavail];
                      int bytesRead = m_client.Receive(buffer,
                                  bytesavail, SocketFlags.None);
                      RxMsgEventArgs args =
                          new RxMsgEventArgs(buffer);
                      MessageRx(args);
                  }
              }
              catch (Exception)
              {
                  break;
              }
        }
        running = false;
    }

    private void MessageRx(RxMsgEventArgs args)
    {
        if (OnMessageRx != null)
        {
            OnMessageRx(this, args);
        }
```

```
        }

        public void SendMessage(string message)
        {
            byte[] buffer = Encoding.UTF8.GetBytes(message);
            m_client.Send(buffer);
            if ( !running)
            {
                rxThread.Start();
            }
        }

        public void Close()
        {
            m_client.Close();
        }
    }
}
```

As you can see, this code is very similar to our server code. We pass into the constructor the server's IP address and the port to use. The IP address is passed as a string in the form 192.168.1.140. We convert that string into an **IPAddress** class instance, using the static class function **IPAddress.Parse**. We use this IP address and port number to create an endpoint for the server. We create a new socket and also set up our poll timeout. The function **Connect()** will use the server endpoint we created to get our client socket to request a connection from the server, using the **Socket** connect function **m_client.Connect()**. We then create a thread to handle the polling of our socket for received messages. Our polling thread is a slightly modified version of our server session thread. It polls the socket for new messages, and if one is received, the message is read into a buffer and an **OnMessageRx** event is fired, with the message buffer.

To send a message to the server we use a **SendMessage()** function, which takes the message buffer and calls **Socket.Send**, as in the server code. We need a delegate and event arguments defined for our public **OnMessageRx** event. The **EventArgs** are simpler than what we used for the server, because they contain only the received message buffer.

Add a new class to your project called **RXMsgEventArgs.cs** and add the following code to define the **EventArgs** and event delegate:

```
using System;

namespace TCPClient
{
    /// <summary>
    /// Rx Message event args
    /// </summary>
    public class RxMsgEventArgs:EventArgs
    {
        public byte[] Message { get; private set; }

        public RxMsgEventArgs(byte[] message)
        {
            Message = message;
        }
```

```
        }

        public delegate void MsgRxDelegate(object sender, RxMsgEventArgs e);
}
```

This code just uses API calls that are available in the Micro Framework, which is a subset of the desktop API, so all calls are available in the desktop .NET. We will see how we can use this class in a Micro Framework application later.

We have our client class, so now we need to add the code to our Program.cs file to use it.

We will use a simple console application to start our client, connect to the device server, and then enter a message to send using the keyboard. The server will generate a response to the message, which we will display in the console window. The application is very simple and does no real handling of error conditions, such as loss of connection to the server.

Modify your **Program.cs** class as follows:

```
using System;
using System.Text;
using System.Threading;

namespace TCPClient
{
    class Program
    {
        private const string SERVER_ADDRESS = "192.168.1.140";
        private const int SERVER_PORT = 1000;

        static void Main(string[] args)
        {
            Client myclient = new Client(SERVER_ADDRESS,SERVER_PORT);
            myclient.OnMessageRx += new MsgRxDelegate(myclient_
            OnMessageRx);
            Console.Write("Connecting... ");

            myclient.Connect();
            Console.WriteLine("Connected\n");
            {
                while (true)
                {
                    Console.Write("Enter a string and press ENTER
                        (empty string to exit): ");

                    string message = Console.ReadLine();
                    if (string.IsNullOrEmpty(message))
                        break;

                    byte[] data = Encoding.Default.GetBytes(message);
                    Console.WriteLine("Sending... ");

                    myclient.SendMessage(message);

                    //pause to give response time to arrive
                    Thread.Sleep(200);
```

```
            }
        }
        myclient.Close();
    }

    static void myclient_OnMessageRx(object sender, RxMsgEventArgs e)
    {
        Console.WriteLine("Response: " +
            Encoding.Default.GetString(e.Message, 0, e.Message.
                Length));
        Console.WriteLine();
    }
  }
}
```

The code will create our client class and set the server IP address and port (make them the same as the settings on your device). It will then call our client connect method and wait for a message to be typed on the keyboard. When a message is typed in, it is sent to the server, and if a response is received, it is sent to the console display. We connect an event handler to the client **OnMessageRx** event to do this.

Now that we have a test client, we can try out our device server application.

1. Open the Server device application in Visual Studio, and connect up your device to the PC using USB, and connect the device to your Ethernet network.

2. Start a debug session in Visual Studio for the device. We will run the device in debug mode from Visual Studio so we can monitor the text debug statements.

3. Navigate to the Bin directory of your client project, and in the Debug (or Release, if you built in release mode) directory, you will find the TCPClient.exe you just built. Double-click to run this Windows console application.

TIP *You can open two instances of Visual Studio. In one you can open the device project and connect, deploy, and debug your device application over USB. In the other instance of Visual Studio, you can open the console client project and run this from the debugger. This way, you can see the debug data from both applications, set breakpoints in either the client or server, and so on.*

---

4. When the Client application starts, it will request a connection to our device server. In the Output window of Visual Studio, you will see the debug text from the device: *Connection to Client*. The client has requested a connection from the server, and the server has responded.

    In the Client Console window, type a test string and press ENTER. This string will be sent as a message and the server session will respond. When the message is received by the server session, details of the message are sent to the Output window as debug text. The response from the server session is displayed in the client console window.

5. If we close our Client application, we disconnect from the session. The server session will detect this and close the session socket.

Using Code in Desktop and Micro Framework

Our client class was used in a desktop test application; however, we used only API calls that are available in the Micro Framework. We have to make only two minor changes to be able to use the client and **RxMsgEventArgs** in a Micro Framework or Gadgeteer application. Both of these are to done with namespaces (in the using statements section).

In the **Client.cs** class we use some **Debug.Print** statements. In the desktop .NET build, this is in the **System.Diagnostics** namespace, but in the Micro Framework this is in the **Microsoft.Spot** namespace. In a C# file, we can have using statements that are not used, and they are just ignored during the build process. So this allows us to put both the desktop and the Micro Framework using statements in the code. If you add a **using Microsoft.Spot;** statement to the Client.cs file, it will compile in both .NET project types.

In a similar fashion, the **RxMsgEventArgs** class inherits from the **EventArgs** class. In the desktop .NET, this is in the system namespace, but in Micro Framework it is in the **Microsoft.Spot** namespace. Adding a **using Microsoft.Spot;** to the **RxMsgEventArgs** class will allow it to compile in both .NET project types. You can now use the client class in a Micro Framework application.

Web-Connected Devices

The Gadgeteer Framework supplies a web server and a web client. These allow you create web-connected devices very simply.

Web Server

The Gadgeteer **WebServer** class allows your device to become a web server, responding to requests from browsers and so on. The web server is basically a TCP/IP server with HTTP-handling abilities. The Gadgeteer web server defines a class called a **WebEvent**. The web event has a pathname associated with it, which will match the path part of an HTTP request. The **WebEvent** has an event **WebEventReceived**, which is fired by the web server when an HTTP request is received with a matching path. The event contains the HTTP data and a mechanism to send a response back to the caller. The URL used to access the device is the device IP address followed by a path reference—for example http://192.169.1.140/*home*, where *home* is our path reference. We create a **WebEvent** and pass in the identifier home. The web server will now add the **WebEvent** to its collection of events. When it receives an HTTP request, it will look for a matching **WebEvent**. When a match is found, it triggers the **WebEventReceived** event for that **WebEvent**. The application handler for the event can now implement the required behavior for the request.

The received HTTP request can also pass in parameter variables in addition to the path. The parameter data consists of name/value pairs. Multiple parameters can be passed. Parameters are defined as [*parameterName*] = [*parameter value*]. A list of parameters are separated by the ampersand (&) symbol. For example, if our board has two LEDs and we want to set their on/off states, we can have two parameters called **Led1** and **Led2**. We set up a **WebEvent** with a path of **LedControl**. Our HTTP request to turn **led1** on and **led2** off will be

```
http://192.168.1.140/LedControl?Led1=on&Led2=off
```

where 192.168.1.140 is the IP address of your device. We will see in our example project how we access these parameters from the **WebEvent** handler.

The **WebEvent** handler allows us to return a response. This response is in the form of a byte array buffer and a standard MIME (an Internet media type) type. This is a standard definition for the type of data being sent, so the browser knows how to handle the data. The **Responder** class used to return the response provides several overloaded methods that will automatically set the correct type for you, or you can set it explicitly.

The current automatically set types are images (JPEG, BMP, GIF), audio MP3, and text. (Why we have audio MP3 as an overloaded method is a mystery to me, because MP3 is not a supported function of the Micro Framework!)

Our web server example project will have a home page, which will supply a very basic HTML page with two links on it, enabling the LED on the mainboard to be turned on or off. When one of the links is clicked in the browser, a new HTTP request will be sent, passing in the LED on/off state as a parameter. We will have a **WebEvent** for the home page and for setting the state of the LED.

Let's see how to control functions on the device using a web browser and how to use the web server and set up web events.

Create a new Gadgeteer project in Visual Studio. I am going to use the Mountaineer board with Ethernet again. Add a mainboard and matching Ethernet module.

Add a new class called **WebApp.cs**. We will add a **WebServer** and configure our two **WebEvents** here. The hardware we are going to be controlling is the debug LED on the mainboard. To access this, we need to have the instance of the mainboard passed into our **WebApp** class. Copy the following code into the **WebApp** class:

```
using System;
using Gadgeteer;
using Microsoft.SPOT;
using Gadgeteer.Networking;

namespace WebServer1
{
    /// <summary>
    /// This is our sample web server app project
    /// </summary>
    public class WebApp
    {
        private WebEvent home;
        private WebEvent webEventDebugLed;

        private Mainboard m_mainboard;
        private string m_ipAddress = "0.0.0.0";

        public WebApp(Mainboard mainboard)
        {
            m_mainboard = mainboard;
            // create our web events
            InitWebEvents();
        }

        private void InitWebEvents()
        {
            home = WebServer.SetupWebEvent("home");
```

```
        home.WebEventReceived +=
            new WebEvent.ReceivedWebEventHandler
                (home_WebEventReceived);

        webEventDebugLed = WebServer.SetupWebEvent("DebugLed");
        webEventDebugLed.WebEventReceived +=
            new WebEvent.ReceivedWebEventHandler
                (webEventDebugLed_WER);
    }

    /// <summary>
    /// This is our 'home' page
    /// Sets up links to turn the
    /// mainboard debug led on or off
    /// </summary>
    /// <param name="path"></param>
    /// <param name="method"></param>
    /// <param name="responder"></param>
    void home_WebEventReceived(string path, WebServer.
                                HttpMethod method,
                                Responder responder)
    {
        responder.Respond(
            "<html><p>Led on <a href=\"http://"+ m_ipAddress +
            "/DebugLed?Led=on\">"
            + "Turn on debug led</a></p>"
            +"<p>Led off <a href=\"http://" + m_ipAddress +
            "/DebugLed?Led=off\">"
            +"Turn off debug led</a></p>"
        );
    }

    /// <summary>
    /// Handler for our led on/off web event.
    /// Called when link from 'home' page clicked
    /// </summary>
    /// <param name="path"></param>
    /// <param name="method"></param>
    /// <param name="responder"></param>
    void webEventDebugLed_WER(string path,
                                WebServer.HttpMethod method,
                                Responder responder)
    {
        bool ledOn = false;
        ledOn = responder.UrlParameters["Led"].ToString() == "on";
        m_mainboard.SetDebugLED(ledOn);
    }

    /// <summary>
    /// Start the server
    /// </summary>
    /// <param name="IPAddress"></param>
    public void StartServer(string IPAddress)
    {
        m_ipAddress = IPAddress;
```

```
        try
        {
            WebServer.StartLocalServer(IPAddress, 80);
            Debug.Print("Server Started :" + IPAddress);
            Debug.Print("Home page http://" + IPAddress + "/home");
        }
        catch (Exception exc)
        {
            // error if unable to start server
            Debug.Print("Unable to start server : no connection");
        }
    }
  }
}
```

In the constructor, we pass in the mainboard reference, so we have access to the debug LED. We then create and set up our two **WebEvents**.

The home event will return a simple HTML page, with two links to turn our debug LED on and off. These will call the **DebugLed** event and pass in a parameter **led**, which will be given a value of either **on** or **off**, depending on which link is used. We attach event handlers to each web event's **WebEventRecieved** event.

The handler for the home event creates a simple HTML string defining the links, which is returned using the **Responder** passed in by the event. The handler for the **DebugLed** event will get the URL parameter **Led** from the responder and test whether the value is equal to **on**. If it is, the mainboard debug LED is turned on. If not, then the mainboard debug LED is turned off.

Our final function is **StartServer()**. We pass in the IP address of the server (the device's Ethernet IP address) and call the **Webserver Start** function. We also need to pass in the port for the server to listen on. We have used the standard HTTP port 80. We print out the URL of the home page to the debug Output window.

This shows the simplicity of the Gadgeteer web server. The complete sequence is to define your web events, add them to the server, implement the handlers for the web events, and then start the server.

As the web server (such as our TCP/IP Server example) just sits in the background, we have to do very little to use it in our application. Add the following code to the Program.cs file:

```
using Microsoft.SPOT;

namespace WebServer1
{
    public partial class Program
    {
        private WebApp m_webApp;
        // This method is run when the mainboard is powered up or reset.
        void ProgramStarted()
        {
            m_webApp = new WebApp(Mainboard);

            string IPAddress =   ethernetForMountaineerEth.
                                 NetworkSettings.IPAddress;
```

```
        m_webApp.StartServer(IPAddress);
        // Use Debug.Print to show messages in Visual Studio's
        // "Output" window during debugging.
        Debug.Print("Program Started");
    }
  }
}
```

In the main application we create an instance of our **WebApp**, passing in the mainboard. We get the device IP address from the Ethernet module and call the **StartServer()** function, passing in our IP address. As before, the code is expecting the network connection to be working. If it is not, then we catch the error in our **StartServer** function.

In debug mode, deploy and run the application to your device. When the application is running, the web server home page URL link is displayed in the Output window of Visual Studio. It will look something like Figure 11-3.

It shows the URL of our home page. We can either type this URL into our PC desktop browser or we can navigate to it using Visual Studio. If you CTRL-click the URL in the Output window, Visual Studio will open a browser page to the address—see Figure 11-4.

If you click the Turn On Debug Led link, the debug LED on your mainboard will be turned on; if you click Turn Off Debug Led, the LED will be turned off.

The Gadgeteer **WebServer** will return a default test page if there is no matching **WebEvent** registered against the URL. If you type **http://***your device ip address* into your browser, you will get the default test page returned. It says, "Hey, it works."

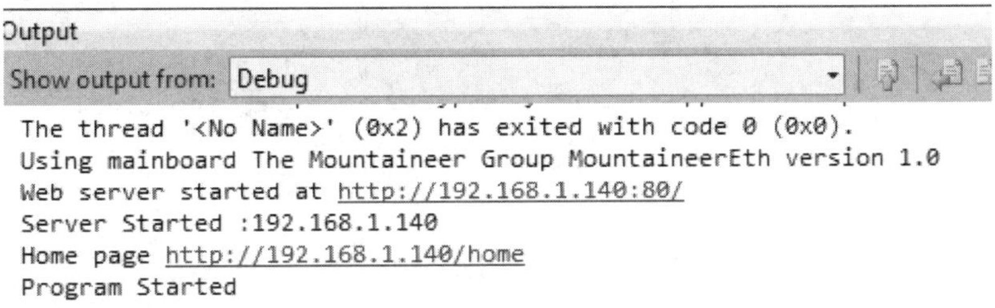

FIGURE 11-3 Project Output window

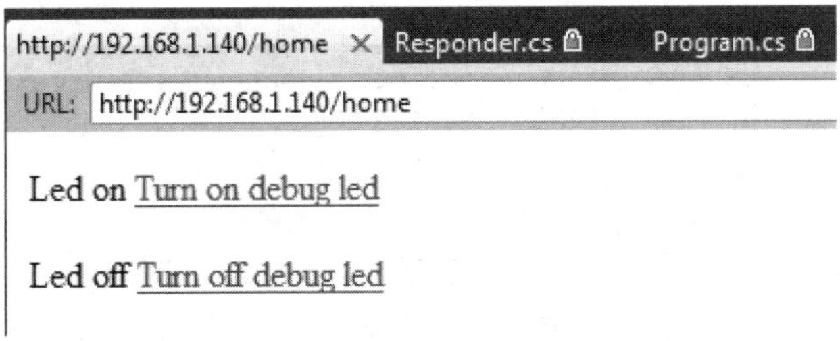

FIGURE 11-4 Web page

Web Client

The Gadgeteer Framework also supplies a **WebClient**. This will allow you to access an external web server. Once again, the basic operation is very simple. A call to the **Client GetFromWeb(*url*)**, passing in the site URL, will return a **HTTPRequest** instance. The **HTTPRequest** has an event **ResponseReceived**. When the URL request receives a response from the external server, this event is fired. Attach a handler to the **ResponseReceived** event to get the data returned by the server. The URL I have used is for one of my web sites and returns a simple text file with the release notes of the NANO board. (By the time you read this, the URL might not exist anymore!)

The following code shows the principles of using the **WebClient**.

```
using Gadgeteer.Networking;
using Microsoft.SPOT;
using GT = Gadgeteer;
using Timer = Gadgeteer.Timer;

namespace WebClient
{
    public partial class Program
    {
        // This method is run when the mainboard is powered up or reset.
        void ProgramStarted()
        {

            GT.Timer timer = new GT.Timer(200,Timer.BehaviorType.
                RunOnce);
            timer.Tick += new Timer.TickEventHandler(timer_Tick);
            timer.Start();

            // Use Debug.Print to show messages in Visual Studio's
            // "Output" window during debugging.
            Debug.Print("Program Started");
        }

        void timer_Tick(Timer timer)
        {
            HttpRequest request;
            request =
            Gadgeteer.Networking.WebClient.GetFromWeb(
            "http://gadgeteerguy.com/Portals/0/SytechFirmware/
                Release4.1.40912.txt");

            request.ResponseReceived +=
                new HttpRequest.ResponseHandler(request_
                    ResponseReceived);

        }

        void request_ResponseReceived(HttpRequest sender, HttpResponse
            response)
        {
            string resTypr = response.ContentType;
            string code = response.StatusCode;
```

```
        int length = response.Text.Length;
        string sample = response.Text.Substring(0, 20);
        Debug.Print("Sample rx file:" + sample);
    }
  }
}
```

In the **ResponseReceived** handler, we print a sample of the text file returned to the debug window.

Additional Micro Framework Network Support

The Micro Framework is not limited to simple TCP/IP-based communications. It also supports User Datagram Protocol (UDP), which is a simpler, faster protocol than TCP/IP as it is a connectionless protocol. This means that it does not have to request a connection from the destination server, as with TCP/IP; it just sends the message with a destination address. As a result, it is also less reliable than TCP/IP because it is a "fire-and-forget" type protocol. When you send a message, you do not know whether it arrived and it is not guaranteed to arrive; however, in most cases, it does!

The Micro Framework also supports UDP broadcast messages and multicast messages. Multicast messages support a subscription/publication scenario. When you send the message (publish), you send it to a multicast address group. Any device interested in receiving these messages will subscribe to this address group. The table of physical device IP addresses subscribing to a multicast group is held by a router. This means the publisher can publish a message to a group and not be aware of the IP addresses of devices that want to receive the messages. It also means devices can subscribe to receive messages from a group, without being aware of the IP address of the publisher (or publishers, as any device can publish to the group address).

If you ever get the chance to ride a tram in Melbourne, Australia, you will see several ticket validation machines around the tram. Each of these machines is sending and receiving messages to the others using UDP multicast messages. Each of the group addresses that are used contains different data, such as which stop the tram is approaching, the GPS position, and so on. Any device can subscribe to the type of messages it requires data from, without being aware of the IP address of the publisher. This is an example of a "publisher – subscriber" messaging system, using UDP multicast messages as the transport. The key principle is that any device can publish messages, associated with a topic, such as GPS position. The publisher is unaware of the devices listening (subscriber) to the message types. Likewise, any device interested in receiving GPS position messages is not aware of the source of the message (the publisher). The only common information needed is the group address (UDP multicast address), of the GPS position messages. Any device that publishes GPS position messages sends to this address. Any device that subscribes to the messages subscribes only to this group address. The router will maintain a table of all devices subscribing to a group address and route any UDP message sent to this address to the devices.

The Micro Framework also supports Secure Socket Layer (SSL). However, the encryption assemblies are relatively large and the support for this function may not be available in some mainboard devices with smaller Flash memory footprints. For more details on using SSL, refer to the MSDN documentation.

Summary

In this chapter, we have covered the basic principles of using sockets to send and receive data over a network. We demonstrated a simple TCP/IP server and a TCP/IP client. We also showed how code written for the Micro Framework can be used in desktop applications.

We looked at the Gadgeteer-supplied **WebServer** and **WebClient** libraries, showing how we can design connected devices using web access.

PART III

Build Your Own .NET Gadgeteer Hardware

CHAPTER 12

Designing Gadgeteer Modules and Mainboards

The guidelines for designing Gadgeteer hardware are included in two documents, available to download from the Microsoft .NET Gadgeteer CodePlex site. The .NET Gadgeteer Mainboard Builder's Guide and the .NET Gadgeteer Module Builder's Guide can be downloaded from http://gadgeteer.codeplex.com/releases/view/72208. These two documents define the mechanical, firmware, and protocol guidelines for mainboards and modules.

The guides include some critical definitions that must be adhered to in addition to some less critical definitions. How strictly you follow the guidelines depends on how you plan to use your hardware. If you intend to sell it publicly, you need to follow the majority of the guidelines to maintain the interoperability aspect of Gadgeteer (the ability for any hardware to work with any hardware).

Here are some of the more critical aspects of the design guide:

- Physical connector sockets and cables

- Pin functions for socket types, not deviating from definitions

- Firmware inheriting from base Gadgeteer classes

If you don't use the same physical connectors and standard cables for your design, other Gadgeteer hardware will not be able to be connected to it. If the functionality of the sockets on your hardware does not follow the standard definition, other Gadgeteer modules cannot connect correctly with them.

Less important aspects of the design specification are the cosmetic items, such as text size and the color of the printed circuit board (PCB) (possibly with one exception: a module that supplies power needs to be obviously marked, so the convention of making these boards red is currently used).

The mechanical specifications are common to all boards. The key points include the specification that all mounting holes are on a 5mm grid, the size of the mounting hole, and the position of the mounting hole with reference to a corner. It is recommended that the corners of the PCBs be rounded. The mechanical constraints are less critical than the electrical constraints. If your board is an odd size and shape, but you follow the electrical constraints, it will still work with other Gadgeteer components.

If the firmware for your mainboard or module does not inherit from the correct base Gadgeteer class, the Gadgeteer core will not be able to integrate your module into a Gadgeteer application.

To simplify the firmware driver process, you can download three builder project templates from the Gadgeteer CodePlex site.

Download the Windows Installer (MSI) package and double-click the file to install these additional project templates into Visual Studio. After installation you will see three new templates in the Gadgeteer Projects page—Mainboard, Module, and Kit—as shown in Figure 12-1.

The Mainboard template is used to create a new mainboard firmware package. The Module template is for modules. The Kit template is for creating a composite install package for a number of modules and optionally a mainboard. This install package will install all the required firmware and design files for your modules and mainboard to a user's PC.

The Mainboard and Module templates will create a new solution with a number of projects. The first project will be used for building MSI install packages for each version of Micro Framework supported. To build an installer project, you must have Windows Installer XML (WiX) version 3.5 or later installed. WiX is an open source add-on for Visual Studio that is used to create standard Windows install packages. It can be downloaded from http://wix.codeplex.com/releases/view/60102.

The template will set up the WiX MSI project for you. You just need to add some further information and list the files you need in the install package. The template will do most of the work for you.

You'll also find a project for each version of Micro Framework you want to support. The project names clearly show the Framework version. The code to implement the driver for your mainboard or module is implemented in these project files. A module or mainboard class for your device is created for you, inheriting from the required Gadgeteer library class. You need to modify this class to generate your firmware driver.

The MSI project will create MSI install packages and merge packages for your module or mainboard. The merge packages can be used by an MSI installer package to create an MSI with several installs. This is the purpose of the Kit template. This will create an MSI package using a number of merge packages for modules and a mainboard. The merge modules are created by the individual install projects you created for each module. The Kit project references the merge files from each of your install projects.

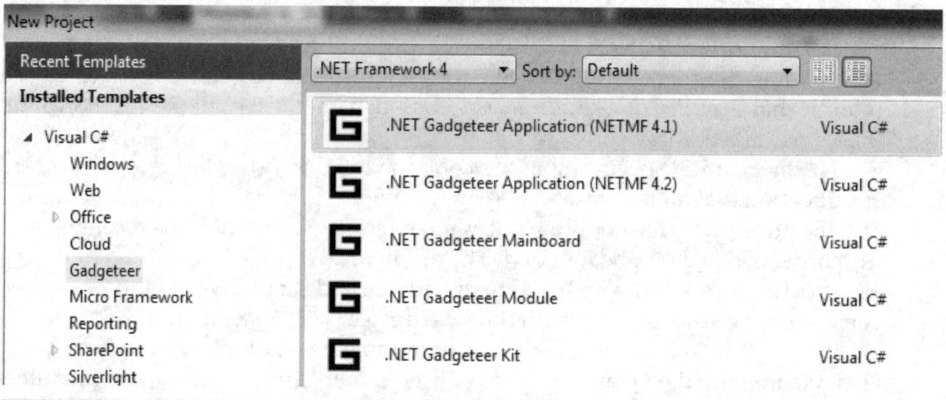

FIGURE 12-1 Visual Studio builder templates

Modules

Module design is simpler and more common than a mainboard design. The module will have one or more Gadgeteer sockets for the physical interface to a mainboard. Usually the interface can be completed with just one socket. The firmware driver for your module needs to derive or inherit from the Gadgeteer **Module** class. The **Module** class is fairly simple and adds the module to the Gadgeteer application module collection.

The second part of your module firmware will normally be one or more Gadgeteer **Interface** classes. These interface classes define the most common standard Gadgeteer socket interface functions such as SPI, Serial, Inter-Integrated Circuit (I2C), and so on. You then add code to implement the specifics of your new module interface.

The best examples of how to write module firmware are on the Gadgeteer CodePlex site. Browse to the Source Code tab. This will open the complete source tree for the latest version of the Gadgeteer source code. Expand the Main tree on the left side of the page, and click the Modules section. You will see the various manufacturers' code sections for modules. The MSR section contains sample module firmware, written by Microsoft Research. For an example of an I2C module, look at the code in the Accelerometer section.

A Simple Custom Prototype Module

It is a relatively simple task to create your own Gadgeteer module from an existing sensor breakout board. In this instance, you need a sensor that is not currently available as an OEM module, but is available as an OEM sensor board (non-Gadgeteer), from a manufacturer such as SparkFun. SparkFun is a source of low-cost hobby and experimenter electronic parts and modules, available at www.sparkfun.com.

For instance, you require a 12-key capacitive touch keyboard for your project. SparkFun offers a breakout board with an I2C interface that supplies this functionality (SparkFun part number SEN-09695). The module has two sets of connections: one for the control interface and the 12 driver outputs to your capacitive key pads. The control interface to the board requires +3V3, Ground, I2C Data, I2C Clock, and an Interrupt signal. You connect up your own custom capacitive keyboard touch pads, laid out on a PCB, to the touch pad driver outputs of the module. A Gadgeteer I (I2C) socket can supply all the required control interface signals.

A prototype of your module can be simply built using one of the available Gadgeteer breakout or expansion modules. These supply one or two Gadgeteer sockets, with the 10 pins connected to a 10-pin header, usually a 0.1-pitch header. This allows you to connect the SparkFun breakout board interface connections to the expansion module header to create a standard Gadgeteer module interface to a mainboard.

Referring to the Gadgeteer Module Builder's Guide, you will see the pin-out required for an I (I2C) socket. Pin 3 is defined as a GPIO pin, with interrupt capability. Connect the Interrupt pin from the SparkFun breakout board to this pin; connect the I2C data and clock lines to pins 8 and 9, respectively; and get +3v3 from pin 1 and ground from pin 10. You now have a Gadgeteer physical interface to your prototype module.

We will now look at the principles of generating the module driver firmware. We will use the project templates from release x.x.x.600 of the Gadgeteer libraries, as this supports both Micro Framework versions 4.1 and 4.2.

Using the Module Project Template

The Module Template will ask for the manufacturer name and the versions for which Micro Framework module support is required. Each version of Micro Framework requires its own firmware driver, using the core libraries built for that version. The template will create a new solution with a number of projects. The first project is specifically for building the MSI install packages using WiX.

1. Create a new project in Visual Studio using the Gadgeteer Module template.

2. In the Module Project Creator dialog, you'll need to enter some general information about your new module. The Module Name is the name of the project. In our example, it's **ProtoModule**.

3. For the Manufacturer Full Name, enter this as a full name and also a safe name with no spaces or punctuation. This example uses Sytech Designs Ltd as the full name and Sytech as the safe name.

4. The check boxes determine whether you want 4.1 and/or 4.2 support. Select NETMF 4.2. See Figure 12-2 for the completed dialog.

In our new solution, we have a number of projects. The ProtoModule project is used solely to build the MSI install packages. The other two projects are for the actual module firmware—one for Micro Framework 4.1 and one for Micro Framework 4.2. We are going to support only Micro Framework 4.2 in our example, so we can delete the Micro Framework 4.1 project from the solution. To do this, right-click the project name and select Remove (see the note for details).

NOTE *The Module Template always adds projects for MF4.1 and MF4.2 irrespective of the version support options selected. Build option constants are set to include the required projects in the build process. In our example we selected just MF4.2 support, so only the MF4.2 project will be included in the WiX build.*

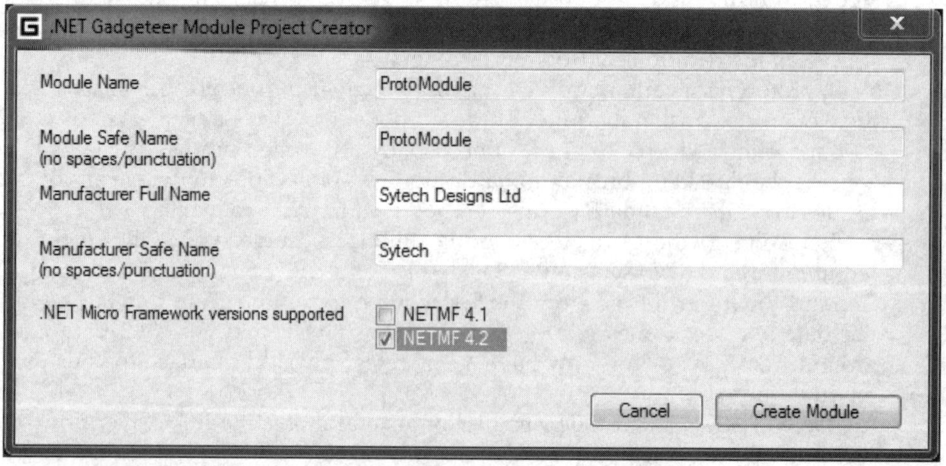

FIGURE 12-2 Module Project Creator dialog

The template will generate a stub class for your module firmware. This is the ProtoModule_42.cs file in our ProtoModule_42 project. The template will add an interrupt interface connected to pin 3 and implement button functionality to this interface. We require an interrupt interface for our board interrupt function, but not the button code. We also need an I2C interface for our control connection to our breakout board, so we will add this. The initial code is modified as shown in the following code listing. (I have deleted all the template comments to shorten the listing.)

```
using Microsoft.SPOT;
using GT = Gadgeteer;
using GTM = Gadgeteer.Modules;
using GTI = Gadgeteer.Interfaces;

namespace Gadgeteer.Modules.Sytech
{
    /// <summary>
    /// A ProtoModule module for Microsoft .NET Gadgeteer
    /// </summary>
    public class ProtoModule : GTM.Module
    {
        // our I2C interface
        private GTI.I2CBus i2c;

        // our interrupt interface
        private GTI.InterruptInput input;

        // Note: A constructor summary is auto-generated by the doc
        // builder.
        /// <summary></summary>
        /// <param name="socketNumber">The socket that this module is
        /// plugged in to.</param>
        /// <param name="socketNumberTwo">The second socket that this
        /// module is plugged in to.</param>
        public ProtoModule(int socketNumber, int socketNumberTwo)
        {
            // our module socket
            Socket socket = Socket.GetSocket(socketNumber, true, this,
            null);
            //Validate Socket choice, ensure it's a 'I' socket
            socket.EnsureTypeIsSupported(new char[] {'I' }, this);

            // This creates an GTI.InterruptInput interface, triggered
            // on a negative edge. Fired whenever a key is pressed or
            // released
            this.input = new GTI.InterruptInput(socket,
                                    GT.Socket.Pin.Three,
                                    GTI.GlitchFilterMode.On,
                                    GTI.ResistorMode.PullUp,
                                    GTI.InterruptMode.FallingEdge,
                                    this);

            // This registers a handler for the interrupt event of the
            // interrupt input
```

```
        this.input.Interrupt += new GTI.InterruptInput.
        InterruptEventHandler(this._input_Interrupt);

        // initialize our I2C interface to our module - I2C
        // address is 0x5a
        i2c = new GTI.I2CBus(socket, 0x5a, 100, this);
    }

    /// <summary>
    /// Public event to notify application of
    /// a key change event.
    /// </summary>
    public event KeyChangeDelegate OnKeyChange;

    #region I2C module handling code
    // use the I2C class to access the registers on the breakout
    // board and detect the key presses. Add the control code
    // here....

    private void _input_Interrupt(GTI.InterruptInput input,
                                  bool value)
    {
        // our interrupt handler will go here.
        // The breakout board will trigger this interrupt
        // every time a key is pressed or released.
        // we would then detect the key (read the relevant I2C
        // registers) and fire our public OnKeyChange event...
    }
    #endregion

}

// Dummy event delegate, real one would have custom event args
// with key press data
public delegate void KeyChangeDelegate(object sender,
                                       EventArgs args);
}
```

This is the outline for our module firmware. We need to implement the actual code to set up the keypad chip and read the keypad status when the interrupt is fired. It shows the principles in creating module firmware.

The key points are that the class inherits from the base **Module** class and it uses standard Gadgeteer interfaces—the interrupt input and the I2C interface. When the code is fully implemented, we can test it by adding a test Gadgeteer project. We delete the designer code generated section and add our module initialization by hand—adding our firmware driver project as a reference in our test project. In the main firmware project, there is a readme.txt file. This offers instructions on how to use the template and test the new module driver.

When we have completed our driver firmware implementation and it is tested, we are ready to generate the MSI install package.

One of the key elements in this package is an XML file that defines the capabilities of our module for use by the Gadgeteer GUI designer. We will now examine this XML file and what we need to add to it.

GadgeteerHardware.XML

The GadgeteerHardware.XML file is in the main MSI project. It defines the capabilities of your module for use by the designer GUI. It defines the socket types on your module and their physical location on the board. The default XML file is partially filled in for you, with the manufacturer information, module name, and so on. The rest of the board parameters need to be added to the other sections of the file; they are all fairly well commented with instructions.

A JPEG image of your board should be added to the project (the Image.jpg file). This is an overhead picture of your module, showing the socket. This image is used by the designer when your module is added to a project. In our example, we can omit the picture and leave the default blank picture in place. The designer will add a grayed-out box using the dimensions of our board.

In the XML file, we define the size of the module board (in millimeters) and the x and y offsets to the center of the connector, from the top-left corner of the board.

We need to define the position of our Gadgeteer socket and define the pins used. We will use a dummy board dimension and socket position, due to the nature of our prototype. So in the first section, we will leave the default board size of 22×44. The dimensions are all in millimeters.

The file will list any assemblies required by the module. These will be added as references by the GUI designer. In our example (and most modules), the only required assembly is the actual firmware driver.

The next node in the file defines the socket. Here we define the type of socket (in our case, type I) and the position of the socket on the board. We have put it in the center of the board. Next we define the pins used on the socket. We are only using pin 3 for the interrupt input and the I2C clock and data lines. On the mainboard, the I2C bus is most likely available on a number of sockets, so the clock and data signals are considered shared. We need to let the designer know that it is OK for other sockets to use these pins as well, because we do not need exclusive use of them, so they are marked as shared. Pin 3 is exclusive; only our module can use it—so it is marked as not shared.

I have removed all the other comments and unused lines from the template XML file for clarity. Our module XML file is shown here:

```
<?xml version="1.0" encoding="utf-8" ?>
<GadgeteerDefinitions xmlns="http://schemas.microsoft.com/
Gadgeteer/2011/Hardware">
  <ModuleDefinitions>
    <!-- The Unique ID is auto-generated and does not usually need to
    be modified. -->
    <ModuleDefinition Name="ProtoModule"
              UniqueId="c87a7ec6-6e3c-40de-92ba-7788b916cc4d"
              Manufacturer="Sytech Designs Ltd"
              Description="A ProtoModule module"
              InstanceName="ProtoModule"
              Type="Gadgeteer.Modules.Sytech.ProtoModule"
              ModuleSuppliesPower="false"
              HardwareVersion="1.0"
```

```
                    Image="Resources\Image.jpg"
                    BoardHeight="22"
                    BoardWidth="44"
                    MinimumGadgeteerCoreVersion="2.42.500"
                    HelpUrl=""
                          >
        <Assemblies>
          <!-- This lists the assemblies which provides the API to this
           module -->

          <Assembly MFVersion="4.2" Name="GTM.Sytech.ProtoModule"/>
        </Assemblies>

        <Sockets>

          <!-- Define a I2C socket in the center of the board -->
          <Socket Left="11" Top="22" Orientation="0" ConstructorOrder="1"
           TypesLabel="I">
            <Types>
              <Type>I</Type>
            </Types>
            <Pins>
              <Pin Shared="false">3</Pin>
              <Pin Shared="true">8</Pin>
              <Pin Shared="true">9</Pin>
            </Pins>
          </Socket>
        </Sockets>
      </ModuleDefinition>
    </ModuleDefinitions>
</GadgeteerDefinitions>
```

TIP *If you hover the cursor over a node value in the XML file, IntelliSense will bring up the description of the value and its use.*

Once again, the XML file is fairly well commented with instructions on use, and the readme.txt file also has instructions.

NOTE *In Micro Framework 4.2 templates, the Gadgeteer assemblies were rearranged slightly from Micro Framework 4.1. SPI, Serial, and Web code was put into its own assemblies (DLLs). When a new application template is used, only the base assembly references are added. If you have a module that requires SPI, Serial, or Web functionality, the GadgeteerHardware.xml file has entries to specify this, causing the designer to add references to these assemblies when a module using them is added.*

MSI Generation

The final step is to set up the project to generate the MSI install package. This will package up all the required DLLs and associated files, place them in a directory where they can be accessed by the Gadgeteer GUI designer, and add your DLLs to the assembly

Global Assembly Cache (GAC). This adds them to the .NET tab when you go to add assemblies in a project. Some registry settings are written to make this process work. Everything will be wrapped up into an auto-install package.

Once again the readme.txt file has fairly complete instruction on this process.

For our simple example, the only files required are the firmware project files. These are automatically set up in the WiX project by our builder template. You need to modify the project only if you have other files to add to the package. We will use the default package version number of 1.0.0.0; if you require a different version number, this is set in the common.wxi file, in the node ModuleSoftwareVersion.

To build the install MSI package, we need to change the build mode from Debug to Release. You will find this in the toolbar at the top of Visual Studio. Click Solution in the Explorer window, and then in the drop-down box in the toolbar, select Release. See Figure 12-3.

When the solution is set to Release mode, select BuildSolution by right-clicking the Solution name in Solution Explorer.

The WiX build will take longer than a normal build. When it has finished, there should be no errors. The MSI install package will be in the bin\Release\Installer directory of the main MSI project directory. We now have a new MSI file called ProtoModule.msi.

Close Visual Studio, if it is open. Double-click the ProtoModule.msi file to install it. Answer the usual Windows security questions about install from an unknown vendor. The new package will be installed. The install dialog is shown in Figure 12-4.

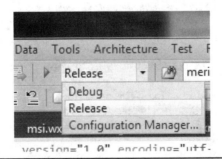

FIGURE 12-3   Select Release mode build

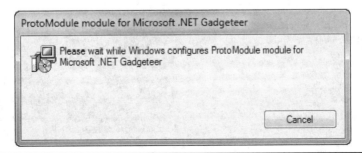

FIGURE 12-4   Install MSI window

After the package is successfully installed, the module is available for use in a new Gadgeteer application. Start Visual Studio and create a new Gadgeteer application. Select a Micro Framework 4.2 project, as our module supports only version 4.2.

A default mainboard will be added by the designer. If you look in the Toolbox, you will see our new module. In our example, the manufacturer's full name was set to Sytech Designs Ltd—so our module is in the Sytech Designs section of the Toolbox. Grab the module and drag it onto the designer. We never supplied an image.jpg, so we have a blank image, set to the size of our module (set in the GadgeteerHardware.xml file). Figure 12-5 shows our module in the designer, connected to an I2C socket on the mainboard.

Note we have set our socket position and socket rotation incorrectly; the socket is not located in the correct place. We can correct this by changing the socket rotation to 90 in our GadgeteerHardware.xml file. This will rotate the socket by 90 degrees, so it is within the bounds of our board. Because this is a dummy module, this is not too important. Any changes will require a rebuild of the project to generate a new MSI. If you want to leave the build number the same (we used 1.0.0.0), then after building you need to use Windows Program Manager to uninstall the package, and then reinstall it—or you need to increment the build number to install over the previous install.

NOTE *The MSI installed module files will be located in \Program Files\Full Company Name \Microsoft .NET Gadgeteer\Modules\ModuleName. You can navigate to this folder to access the GadgeteerHardware.XML file. You can make any final adjustments to socket position, and so on, by hand and save the file. Then reopen Visual Studio and the project to see the effect of the changes immediately. This way, you can fine-tune the positioning of any sockets, and so on, and visually check the result. When you are satisfied with the placement, you can apply the final settings to the XML file in the project and then rebuild the MSI.*

In the new test project, if you inspect the project references, you will see the DLL reference for our new module has been automatically added. If you click on Add Reference in the Solution Explorer and look at the .NET tab, you will see that GTM.Sytech.ProtoModule has been added as a .NET assembly.

To create firmware and an install package for a new module is actually quite a complicated process. But the complexity of generating the required GUI designer interface and the WiX MSI project is simplified by the Module Builder Template. The template project makes creating a basic module package fairly simple.

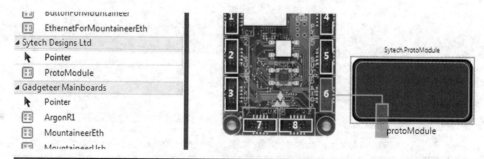

FIGURE 12-5 Using the module in the designer

Armed with this knowledge, you will find creating your own custom modules a reasonably painless process.

Mainboards

The process for creating a custom mainboard driver and install package is basically the same as creating a module. However, the process of designing your own mainboard is much more complex than a simple module. It is unlikely you will ever be in the position of needing to design your own mainboard, because several mainboards on the market offer varying capabilities and costs, and they should fulfill most ordinary requirements.

The critical component required for your own mainboard is always going to be a working Micro Framework port for that hardware. The porting process for a processor is a very advanced topic and beyond the scope of this book.

If you are interested in seeing how the firmware drivers work for a mainboard, you can find several examples on the Gadgeteer CodePlex site. Most manufacturers publish the firmware drivers on CodePlex as open source code.

CHAPTER 13

Turning Your Gadgeteer Prototype Into a Product

U sing a mainboard, some modules for the hardware, and the Gadgeteer libraries, you have managed to prototype the hardware and software for your design. Now you want to turn this prototype into a real product.

If the production volume is small, then using the Gadgeteer OEM modules and mainboards is the way to go. If the production volume will be higher, then a custom-designed board that incorporates the mainboard and modules should be seriously considered. We will examine the costs of using OEM boards and a custom board for a relatively simple design. You will see that in general a custom board is not economical until your production quantity is more than 100 units.

Use Modules or Design a Custom Board

The key factor is how many units you want to make. If you plan to build a small number of units, such as 10, then the most likely solution is to stay with the OEM mainboard and modules that you are using. However, if you need to make 500 units, this solution is unlikely to be the most practical or economical. In this case, you'll need to design or commission a custom printed circuit board (PCB) that incorporates your mainboard and modules.

Most Gadgeteer mainboard and module designs are open source, so you can use the physical circuit designs (and the firmware) without paying license fees. If, for example, your design uses a mainboard and four modules, this would involve five PCBs and at least four cables and connections. The costs associated with manufacturing a custom PCB are mainly setup and handling charges, because the material is not a significant part of the equation. So, for example, it doesn't cost much more to make a PCB that is 20mm-by-20mm than a PCB that is 100mm-by-100mm. Likewise, if you put multiple copies of your board onto a single panel, the manufacturing cost can be reduced even more. It generally costs less to have a PCB fabrication house manufacture 10 panels, with five boards on each panel (total 50 boards), than to make 50 identical PCBs as a single board.

It is inevitable that the design will use surface-mount components (surface-mount technology, or SMT). These components are mounted on pads on the board, as opposed to plated through hole (PTH) components, which have legs or pins for each component's connection. It is possible to hand-solder SMT components, but this is practical only for very low volumes, and it is an extremely labor-intensive and skilled process. The normal method is to have an assembly house assemble the boards for you, where specialized (and very expensive) computer numerical control (CNC) "pick-and-place" machines are used to "pick" the component from a component bin and "place" it precisely on the pads on the board. Once again, the main cost is setup, because the machine needs to be programmed to know which components to use and where to put them on your board. A board arranged into panels is less expensive to assemble. After one circuit has been programmed, this can be simply "step and repeated" for the other boards on the panel.

So let's compare the current costs of our example product, using a mainboard and four modules:

- Mainboard: $29.95, Cortex M4 board
- Module 1: $9.95, power supply and USB device
- Module 2: $19.95, USB virtual serial port
- Module 3: $12.90, accelerometer
- Module 4: $4.95, button and LED

This is a total of $77.70.

We will compare the costs of 10 units and 100 units, using OEM modules and a custom board.

Using the modules, costing is simple:

- 10 units = $777.00—unit price $77.70
- 100 units = $6993.00—unit price $69.93

You negotiated a discount with the supplier on the 100 units for a volume buy!

For costing the custom board, let's assume the board will be 60mm-by-80mm, and because the design is simple, we will use a two-layer board:

- PCB design work: approximately $1200
- PCB fabrication:
  - 10 boards = $221 ($22.10 each)
  - 100 boards = $484 ($4.84 each)
- PCB assembly:
  - 10 boards = $640
  - 100 boards = $1600
- Parts cost:
  - 10 boards = $24.75
  - 100 boards = $1950.00
- Total cost of a manufactured board:
  - 10 boards = $2308.50 ($230.80 per board)
  - 100 boards = $5234.00 ($52.34 per board)

The results show that we need to manufacture about 100 boards before a custom board is economical. However, a significant one-time cost is the PCB design work. If we make a second batch of 100 boards, these will cost $1200 less than the first batch, a more than 20 percent cost reduction. Clearly, it is not economical to make ten units as a custom design, unless there was some other overriding design factor such as size reduction. In our example of ten units, a custom board is almost four times the cost of the OEM board solution.

TIP *Some mainboards and modules also have Eagle design files for the circuit and PCB. Eagle is a very low cost circuit and PCB design package. Using Eagle, you can design the PCB yourself, using these design files as a starting point, and save the PCB design service fee. Several other low-cost design tools are available, and some will allow you to import Eagle files as well.*

A custom board offers many advantages. You have control over the size and shape of the board, connector placement, and so on. A single board is going to be a lot more reliable than five boards interconnected with cables. But the most important advantage, from a commercial point of view, is that you have total control over the production process, and you are not reliant on third-party products.

NOTE *I would recommend using a design consultancy experienced with Micro Framework hardware and software design to help get your design into production. However, because my company supplies design services, I guess I might be a bit biased here!*

Packaging Your Prototype

You have the idea and have seen a need in the market for your product. Your next major hurdle is the cost of packaging your product to move it from idea to market. The development process is normally fairly expensive. Gadgeteer and OEM modules will allow you to prototype and get to a working proof-of-concept stage for a relative low cost. The main resource you need is your time. A working prototype, correctly presented, can be used to pitch to investors to raise the necessary finances to put your idea into volume production.

Due to their nature, Gadgeteer prototypes can look a bit like a science project—boards and cables everywhere! Being able to package your prototype into a custom enclosure is an important step toward it becoming a product.

The mechanical side of the Gadgeteer specification was designed to make life a little simpler. Interconnection between boards uses the same ribbon cable. The connector polarities are arranged such that a module connects to a mainboard, without the cable twisting. The mounting holes on the mainboards and modules should all be the same diameter (3.2mm diameter, allowing an M3 screw to be used), and the hole spacing will be on a 5mm grid. The holes will be 3.5mm from the edges of the board, and corners will be rounded.

Three-dimensional (3D) printing of an enclosure can be a simple answer to the problem of a custom enclosure.

On the Gadgeteer CodePlex site, you'll find several 3D models for a number of the current Gadgeteer hardware range (44 models at last count). This makes it simpler to use a 3D modeling application to design an enclosure for your prototype. This 3D design can then be used to create a physical enclosure using a 3D printer. Unfortunately, 3D printers and 3D design packages can be extremely expensive, but many companies offer 3D printing services at a reasonable cost. Some of these are Ponoko, Shapeways, and i.materialise. Just type the company names into a search engine to find their web sites.

Some 3D modeling applications are Autodesk 123D, Dassault Systèmes SolidWorks, Autodesk Inventor, and Alibre. Some of the expensive 3D modeling applications offer academic pricing, with a huge discount on the normal price, but you need to be a student and they can't be used for commercial purposes. But 123D is a free application and works with the 3D model files from the Gadgeteer CodePlex site.

Using these low-cost 3D modeling options allows you to do the design work yourself and generate a "first-pass" enclosure. Because you are doing all the work yourself, the cost for this stage of the process is greatly reduced from the conventional step of employing a design consultancy.

Gadgeteer and Micro Framework 4.2

The Gadgeteer library is an extension to the Microsoft .NET Micro Framework. In common with all operating systems, it is constantly evolving and new versions are released. When Gadgeteer was first released, it was built using Micro Framework version 4.1. At the time of writing, the current version of Micro Framework is 4.2 and the Gadgeteer library is at version 2.42.600. This version supports both Micro Framework 4.1 and Micro Framework 4.2. The design environment for both versions is Visual Studio 2010.

When you write an application, all aspects of the code, Micro Framework, and Gadgeteer libraries must be built using the same version of the Micro Framework operating system (SDK version). The mainboard operating system firmware must also use the same version of the Micro Framework.

If your mainboard supports Micro Framework 4.1 QFE1, your Gadgeteer application must be targeted to MF 4.1 QFE1.

If you have the Micro Framework 4.2 SDK installed, support is provided for MF 4.1 and MF 4.2 projects. Gadgeteer version 2.42.600 also supports applications in MF 4.1 and MF 4.2. Project templates are supplied for creating MF 4.1 and MF 4.2 applications. When you select an MF 4.1 or MF4.2 Gadgeteer application template, the design environment in Visual Studio will be set up to use the correct versions of the DLLs. The Micro Framework 4.2 SDK will install two sets of core DLLs: one for 4.1 and one for 4.2. Registry settings are set so the Visual Studio environment knows where to find the correct DLLs.

Gadgeteer MF 4.1 and 4.2 Applications

As mentioned, Gadgeteer version 2.42.600 supports writing MF 4.1 and MF 4.2 applications. The library is slightly different between the two versions. In MF 4.1 applications, the project template will add the complete set of Gadgeteer library DLLs to the project. The main library DLL contains all the Gadgeteer Interface classes, so you deploy all the code library, such as an SPI interface, to your application, even if you never used any modules with an SPI interface. This meant a larger code base was deployed to your device, requiring a larger memory footprint.

With the advent of a number of devices using the STM Cortex M3/M4 processors, the available memory on the hardware was reduced. The advantage of using this hardware is that the mainboard can be a "single chip" solution. The Flash and RAM

Micro Framework Release Versions

Micro Framework 4.2 has two main releases, MF 4.2 QFE1 and MF 4.2 QFE2. MF 4.2 QFE2 is the final version of MF4.2. The main reason for two versions was a potential problem with the Windows USB driver. Some hardware combinations of PC and mainboard threatened the possibility of a serious crash on the desktop machine if the USB connection was terminated under certain conditions. This caused a Blue Screen of Death (BSOD) type crash. This was considered a serious enough problem to require an immediate update version, rather than waiting for the next scheduled version update (MF 4.3). MF 4.2 QFE2 adds base code to allow the USB driver to be a Win USB type driver, rather than a custom USB driver. The original USB driver is also supported. Win USB driver functionality on a PC is run in a protected mode, so it should not be able to crash the core operating system. This gave mainboard hardware manufacturers the option of changing over to the Win USB driver. It is up to the hardware manufacturer which USB driver it incorporates in its firmware port. Some other minor improvements and bug fixes were also incorporated and an additional support function for digital to analog convertors (DACs) was added.

Check with your mainboard manufacturer's documentation to determine what firmware is supported: QFE1 and/or QFE2. Your Micro Framework SDK needs to be the correct QFE version. There are subtle differences in the DLLs; if your mainboard supports QFE1 and you have QFE2 installed (or vice versa), you could experience compatibility problems.

memory are inside the chip. This considerably reduces the cost of the hardware, but the downside is the limit on the amount of available memory.

To take full advantage of these new devices, the MF 4.2 version of the Gadgeteer libraries was rearranged. Some of the interface classes were split out from the main library into their own DLLs. The Ethernet-related code was also put into its own DLLs. The GUI designer was extended to make a new MF 4.2 project reference only a minimum set of the Gadgeteer libraries, making a smaller code base. The module definition functions were extended to allow a module to specify which Gadgeteer libraries it required. When you add a module to the application using the designer, the designer will now add any references to additional DLLs required. This means your code base includes only the library code it requires, making for a smaller final footprint.

MF 4.3 and Visual Studio 2012

The following discussion is based on prerelease information. It is possible that some of these details will change for the final public release.

Every couple of years, Microsoft has a major shake-up with its operating systems and development tools. We are on the verge of one of these events.

With the imminent release of Windows 8, we are also getting new development tools in the shape of a new Visual Studio release, Visual Studio 2012. Coinciding with this will be a new release of the Micro Framework (Micro Framework 4.3) and the SDK integrated with Visual Studio 2012.

There will be an Express version of Visual Studio 2012 that supports Micro Framework and Gadgeteer. In fact, several new versions of Visual Studio 2012 Express

Embedded and Referenced Assemblies

When a firmware image for Micro Framework is built, the manufacturer can choose to embed the managed DLL code in the Flash image or have it downloaded when required by Visual Studio. There are various checks in the Micro Framework core system to ensure that the "managed" part of the code matches the "native" part of the code. Most of the .NET libraries have a managed DLL version and a matching native code section (on Flash). So in general it is usual to embed the managed part of the code in the image. This helps ensure that the correct DLL versions are used.

Some manufacturers are using this feature to help make using the single-chip solutions a bit more efficient. Several .NET DLL libraries (such as graphics) are not needed for every application. If you are not using a display or an SPI display and have no graphics in your application, you don't need the graphics DLLs. So they are not embedding some of the .NET DLLs in the Flash image. This gives you more room for your application image. If your code does use one of the non-embedded DLLs, then when you deploy your application from Visual Studio, it will detect that the DLLs are not in Flash and they will be deployed (downloaded) along with your application. Under most circumstances, this will work fine. But if you have firmware for MF 4.2 QFE1 and build and deploy using MF 4.2 QFE2, the wrong version of the DLL will get downloaded. At best, this will be detected during download or will produce a runtime error, but it can also crash the OS and give you no indication as to the problem.

You can test which DLLs are embedded in Flash using MFDeploy. First, ensure that you have a clean version of your mainboard—that is, no applications downloaded.

1. Set it to this condition by connecting to MFDeploy and clicking the Erase button on the main page.

2. Select the User Application section to delete (not the Firmware option if it is visible). This will delete any user application from your device.

3. From the main toolbar menu, select Plug-in | Debug-Show Device Info. This will write a list of all the DLLs in Flash to the output window.

Note that after you install anything using Visual Studio, these DLLs will also be listed, so you'll need to put the device back into a clean state before this test.

will support different development environments, such as Windows Phone 8 and Microsoft apps.

The Express version supporting Micro Framework will be released shortly after the main versions of Visual Studio 2012.

The Micro Framework 4.3 SDK will support MF 4.1, MF 4.2, and MF 4.3, as will the Gadgeteer library for MF 4.3. It is expected that manufacturers will be releasing MF 4.3 firmware versions for their mainboards fairly quickly.

There are a lot of elements to pull together for this new release involving a lot of work, demonstrating Microsoft's commitment and support of .NET Micro Framework and Gadgeteer.

Index